Mechanochemistry in Materials

Polymer Chemistry Series

Titles in the series:

1: Renewable Resources for Functional Polymers and Biomaterials
2: Molecular Design and Applications of Photofunctional Polymers and Materials
3: Functional Polymers for Nanomedicine
4: Fundamentals of Controlled/Living Radical Polymerization
5: Healable Polymer Systems
6: Thiol-X Chemistries in Polymer and Materials Science
7: Natural Rubber Materials: Volume 1: Blends and IPNs
8: Natural Rubber Materials: Volume 2: Composites and Nanocomposites
9: Conjugated Polymers: A Practical Guide to Synthesis
10: Polymeric Materials with Antimicrobial Activity: From Synthesis to Applications
11: Phosphorus-Based Polymers: From Synthesis to Applications
12: Poly(lactic acid) Science and Technology: Processing, Properties, Additives and Applications
13: Cationic Polymers in Regenerative Medicine
14: Electrospinning: Principles, Practice and Possibilities
15: Glycopolymer Code: Synthesis of Glycopolymers and their Applications
16: Hyperbranched Polymers: Macromolecules in-between Deterministic Linear Chains and Dendrimer Structures
17: Polymer Photovoltaics: Materials, Physics, and Device Engineering
18: Electrical Memory Materials and Devices
19: Nitroxide Mediated Polymerization: From Fundamentals to Applications in Materials Science

20: Polymers for Personal Care Products and Cosmetics
21: Semiconducting Polymers: Controlled Synthesis and Microstructure
22: Bio-inspired Polymers
23: Fluorinated Polymers: Volume 1: Synthesis, Properties, Processing and Simulation
24: Fluorinated Polymers: Volume 2: Applications
25: Miktoarm Star Polymers: From Basics of Branched Architecture to Synthesis, Self-assembly and Applications
26: Mechanochemistry in Materials

How to obtain future titles on publication:
A standing order plan is available for this series. A standing order will bring delivery of each new volume immediately on publication.

For further information please contact:
Book Sales Department, Royal Society of Chemistry, Thomas Graham House, Science Park, Milton Road, Cambridge, CB4 0WF, UK
Telephone: +44 (0)1223 420066, Fax: +44 (0)1223 420247
Email: booksales@rsc.org
Visit our website at www.rsc.org/books

Mechanochemistry in Materials

Edited by

Yoan C. Simon
University of Southern Mississippi, USA
Email: yoan.simon@usm.edu

and

Stephen L. Craig
Duke University, USA
Email: stephen.craig@duke.edu

ROYAL SOCIETY
OF CHEMISTRY

THE QUEEN'S AWARDS
FOR ENTERPRISE:
INTERNATIONAL TRADE
2013

Polymer Chemistry Series No. 26

Print ISBN: 978-1-78262-146-1
PDF eISBN: 978-1-78262-388-5
EPUB eISBN: 978-1-78801-239-3
ISSN: 2044-0790

A catalogue record for this book is available from the British Library

The Royal Society of Chemistry is a charity, registered in England and Wales, Number 207890, and a company incorporated in England by Royal Charter (Registered No. RC000524), registered office: Burlington House, Piccadilly, London W1J 0BA, UK. Telephone: +44 (0) 207 4378 6556.

For further information see our website at www.rsc.org

Printed in the United Kingdom by CPI Group (UK) Ltd, Croydon, CR0 4YY, UK

Preface

Writing a preface is an opportunity to convince the reader (that would be you) of the importance of a monograph and the value of its content, but it is also a chance to frame the theme within a wider, maybe more entertaining, context. It is a central and somewhat stressful task. Conveniently, *stress* is at the center of the present manuscript, more specifically the relationship between stress and (polymeric) materials and the new strategies that have arisen in the past decades to harness mechanical energy and transduce it into useful chemistry in materials.

For many years, materials have been structural: a plush polyurethane foam to make a cushy seat, a glassy polycarbonate for plane windows, a rubbery vulcanized tire to attenuate the asperity of the road. Polymeric materials surround us. The last three decades have seen the unabated development of an ineluctable trend: we now want to impart some "wit" to our materials. Three decades ago a mechanically "smart" material was one whose properties were well suited to its intended use. Materials were designed to withstand a certain load or to allow a certain amount of deformation. The material had to resist or comply. In the intervening years, the moniker of smart materials came to refer to systems that are capable of responding in a predetermined manner to a given stimulus. Examples of smart materials include pH-responsive micelles for drug delivery, programmable shape-memory materials that can twist and tie or even materials that can move left or right depending on the polarization of incident light. It is now clear that mechanical response can be used as a similar stimulus-response mechanism.

Growing up as kids, judo was a somewhat popular sport and while neither of us practiced it one of the elementary tenets of that martial art is the utilization of the force of one's opponent to one's own advantage. One can draw an analogy between mechanochemistry and this creed. Indeed, after many

Polymer Chemistry Series No. 26
Mechanochemistry in Materials
Edited by Yoan C. Simon and Stephen L. Craig

Published by the Royal Society of Chemistry, www.rsc.org

years of designing materials to fight the effect of mechanical action (which thankfully we still do for certain structural components), it is now clear that mechanical force can be used to one's advantage. The present treatise looks at the contribution from various researchers from around the world towards that direction. The range of scientists from different backgrounds is necessary, given the intrinsic interdisciplinary nature of this work. Chemists, surface scientists, materials specialists and engineers alike bring important contributions to the field.

Going back to natural systems, adaptability and feedback loops in living organisms are crucial for their survival. Of particular relevance to our topic of interest are the multiple examples in nature that make use of a graduated or adaptive response to mechanical stimuli such as actin polymerization, stem cell differentiation, or response to an edema, *etc*... Understanding the solutions that Nature has evolved in response to these challenges is therefore a good starting point for the present monograph. The first chapter of this book is therefore concerned with this biological inspiration, taking a look at various instances where Nature needs to respond in mechanical force, looking (in awe) at the sophistication of said systems.

Early in the quest for mechanically responsive systems, it was realized that the conversion of mechanical energy into a useful chemical potential could only be realized through the careful design of functional groups capable of preferentially activating in response to mechanical forces, *ergo* generating a desired chemical function. The second chapter focuses especially on the design of such functional groups, or mechanophores, classifying them according to their ultimate purpose, *e.g.* creation of reactive species, catalytic activation, chromatic change. Scientists worldwide have indeed shown incredible resource to devise and synthesize systems with diverse natures (organic, organometallic), geometries (linear, cyclic) or end goals.

Among the obvious and promising technological outlook for these polymers lies the ability to sense stress. Much like somatosensory cells are able to inform you of a pleasant touch or a painful pinch, it is often crucial to be able to evaluate the stress state of a material, *e.g.* to forestall fatigue or, worse, failure. Optical cues are especially attractive for their sensitivity, spatial and spectral resolution and maybe more importantly their facile observation. Chapter 3 goes over some of the methodologies that have been developed for sensing, including those based on chain scission, aggregation, isomerization, *etc.* These detection strategies require a combined understanding of both chemistry and stress-transfer phenomena.

The latter are explained in greater details in Chapter 4, which focuses on the design principles to effectively transduce mechanical force. In fact, while chemists have found interesting ways to probe new mechanophores through elongational flow, the veritable challenge of mechanochemistry probably lies in the aptitude to transfer as efficiently as possible a macroscopic force to the judiciously designed mechanophores. This section aptly presents the mechanics background to explain transfer of load from the matrix down to the mechanophore level, taking into account experimental conditions, the nature of the matrix and kinetics considerations pertaining to the mechanophore.

Along these lines, it is interesting to evaluate the influence that other chains can have on the overall mechanoresponsiveness, especially *via* the introduction of directional non-covalent interactions. Chapter 5 goes over the recent developments in the field of supramolecular chemistry, rightfully acknowledging that the introduction of strong secondary interactions will have a tremendous impact on the microstructure of a material and, eventually, the ability to transfer stress. Of particular interest to that section are the series of materials that contain supramolecular motifs both in the bulk and as the mechanoresponsive unit.

Understanding what governs force transfer in the bulk is crucial to create materials that respond in a desirable manner to stress. Yet, one easily recognizes that surface and interfacial properties are often as important as, if not more important than, bulk behavior. With that in mind, Chapter 6 highlights some of the most recent findings in the area of polymer brushes and how swelling behaviors and chain conformation have a tremendous impact on coating stability. It goes over some of the compositional considerations that affect the overall resistance of the brush and its attachment to the substrate.

Finally, Chapter 7 highlights the premises of a novel concept in the field of chemo-mechanical transduction converting the energy of an alternating reaction to generate gels that throb autonomously but that are also capable of detecting a force being applied. This adaptability opens up the road for the fabrication of systems that are capable of computing and to respond in an appropriate manner to a given stimulus. As such, the strategies described help move mechanically "smart" materials from being merely obedient, to being capable of adapting their properties to reach endpoints that might not have been specifically designed by the scientist.

It is often acknowledged (and rightfully so) that "a professor is only as good as his students". As editors of this monograph, we gratefully acknowledge that one should extend this to say that a book is only as good as its contributors. We are very fortunate to have convinced some of the leaders in the field to take some of their precious time to contribute to this book. We fully appreciate their patience, involvement and dedication towards the realization of this monograph.

Yoan C. Simon and Stephen L. Craig

Contents

Polymer Chemistry Series No. 26
Mechanochemistry in Materials
Edited by Yoan C. Simon and Stephen L. Craig

Published by the Royal Society of Chemistry, www.rsc.org

CHAPTER 1

Mechanochemistry: Inspiration from Biology

TAMUKA CHIDANGURO, WENGUI WENG AND YOAN C. SIMON*

University of Southern Mississippi, Hattiesburg, MS, USA
*E-mail: yoan.simon@usm.edu

1.1 Introduction and Historical Perspective

Mechanochemistry from the contraction of μηχανικός *mēkhanikos* (mechanic) and χημία *khēmia* (chemistry) is the study of the evolution of the formation and disruption of chemical bonds upon application of an external force. The terminology "mechanochemistry" was first coined by Ostwald as "the coupling of mechanical energy and chemical energy".[1] Unlike electromagnetic force or gravity, mechanical force implies contact. Indeed, for a force to be applied on a given object or organism, it entails a connection or at least transmission of forces through a medium. Mechanical energy, often in the form of applied force, is a lesser-known way to initiate chemical reactions than conventional stimuli (*e.g.* heat, light and electricity). At first glance, it is especially interesting and somewhat intriguing to think of chemistry triggered by forces, as it is not something that is typically taught in your General Chemistry 101 class. However, bond ruptures upon mechanical action surround us from random chain scission as one tears through packaging material or simply as one presses on a surface and senses the mechanical deformation. In prehistory, our ancestors took advantage of mechanochemical phenomena for survival, *e.g.* drilling wood for fire. Theophrastus of Eresus (*ca.* 315 BC),

Polymer Chemistry Series No. 26
Mechanochemistry in Materials
Edited by Yoan C. Simon and Stephen L. Craig

Published by the Royal Society of Chemistry, www.rsc.org

Aristotle's student, first recorded that grinding cinnabar in a copper mortar using a brass pestle could reduce it to mercury.[2] In the 19th century, Faraday and Lea applied sliding and grinding to study the chemical reactions of solid substances.[3]

In many other ways though, for the better part of its existence, humankind has tried to use materials that would minimize their alteration upon application of force. For instance, Damascus steel, which dates back to 900 AD, was a metal of choice for weaponry as it remarkably exhibited superplasticity along with incredible levels of hardness. Interestingly, it is also one of the early examples (though unwittingly) of nanotechnology and mechanochemistry as carbon nanotubes as well as cementite nanowires were proven to form during the forging and annealing of Indian wootz steel. It is indeed believed that the combination of heat and mechanical action along with impurities in the ore (including carburizing wood and leave additives) was responsible for the catalyzing the process responsible for the formation of these nanostructures and the characteristic wavy patterns.[4] The quest for materials that would best resist mechanical constraints has been a technological driving force throughout the Anthropocene and the field of polymer science is no exception to this pursuit. For example, worldwide many groups have been looking at nanocomposites as a source of reinforcement of lightweight materials, with many of these solutions having made it to commercial products in the fields of transportation, construction or even appliances.[5]

From Staudinger to Melville,[6,7] mechanical force was studied early on and the random scission often seen as a foe engendering undesirable chain scission and weakening or worse causing the rupture of polymeric materials. However, in the early 2000s, a new idea emerged in the realm of polymer science.[8] Namely, it became apparent that one could utilize mechanical force to one's advantage rather than combating it. Sensing, repair or self-stiffening were some of the many putative functionalities proposed in the context of mechanochemistry (which have been achieved today and are described in detail in the following chapters). Interestingly and regardless of whether Nature was a source of inspiration for some of the seminal work, it is worth noting that these very functions are ubiquitous in biological systems. Like in many other areas of the sciences, materials researchers recognized early on that one could learn a tremendous amount of information from studying biological materials and understanding how millennia of evolutions have shaped the functionalities of biological systems. Consequently, the present chapter focuses on the mechanochemical strategies developed by Nature to achieve some of the aforementioned functions. It aims to familiarize materials scientists (who may not always have the opportunity to investigate them) with some of the biomechanics and biophysics reports that cover these intricate (and sometimes not entirely understood) processes. This chapter was written from a polymer scientist's perspective and will therefore remain partial and sometimes simplistic. This "editorial line" is substantiated by two main reasons. First, the audience of this book is presumed to be mostly polymer researchers and it would be illusory to condense the complexity of biological mechanotransduction pathways and make it fully accessible in a few pages.

Second, the point of this chapter is rather to inspire materials scientists, and provide them with a general overview of the strategies utilized by living organisms to sense and adapt to mechanical constraints. Consequently, a broad stroke approach seems more adequate as it would appear nonsensical and unachievable to reproduce exactly what Nature does to transduce mechanical forces. Instead, one can draw overarching guiding principles upon which to base his/her reflection and ultimately the design of biomimetic mechanotransducers. Interestingly, there are already several reports of materials which are (wittingly or not) akin to biological systems and, whenever suitable, we will draw a parallel between artificial and biological systems.

1.2 Biomimetism and Rationale for Emulating Mechanotransduction Pathways

Before delving deeper into the intricacies of mechanotransduction, it is worth doing two things: (i) remembering some of the key principles of how to approach biomimetism and (ii) explaining the reason why mechanotransduction is critical for life.

1.2.1 Principles of Biomimetism and Strategies to Implement It

In 1994, a team of researchers led by David Tirrell laid the foundation for the study of materials of biological origin, specifically hierarchically organized structures.[9] Such hierarchy is also present in mechanotransduction schemes and it stands to reason that some of these general ideas developed in their report are readily applicable to mechanotransduction. Tirrell and coworkers highlighted some of the commonalities in materials of biological origin, which are worth recalling. Particularly, we will highlight the properties as they pertain to mechanotransduction.

Evolutionary engineering has advanced by means of an iterative process whereby the structure was refined slowly at each generation. As will become apparent, transduction pathways often find a great deal of commonalities, probably originating from shared ancestral strategies. Many of the elementary units (evidently nuclear bases and amino acids but not only, *e.g.* microfibrils) are recurrent such that function often comes from specific assembly rather than the building blocks themselves. This means for instance that orientation control plays a critical role in determining the assembly responsiveness. The latter is usually adapted and gradual, and varies according to the task performed (*e.g.* slow adaptation *vs.* imminent danger). Shape plays a pivotal role in dictating and modulating the response and shape complexity ensures the right response to the right cue. Since the (mechanotransduction) tasks often prove repetitive, resistance, durability and resiliency are essential. These properties are often predicated upon subjacent principles of reversibility or better yet of mendability. To achieve these paramount functions, Nature often capitalizes on the utilization of non-covalent forces as well as out-of-equilibrium dynamics.[10] While the former are readily achievable and have been realized in the field of mechanochemistry (see Chapter 5), the latter is harder to implement in synthetic systems as dissipative structures are still a research curiosity. It is interesting to note that many of the biological systems involved in mechanotransduction in cells are either in kinetically trapped states (*e.g.* folded proteins, cell membranes) or simply in a dissipative state (*e.g.* the formation/rearrangement of the cytoskeleton to accommodate deformations).

Also of interest is the realization that the mechanical coupling usually occurs between objects of disparate sizes. Interestingly, the interactions maintaining these objects together are often weak and their chemical and thermal stabilities moderate. In that sense, artificial systems are likely to display superior performance and will therefore be more adapted to some of the harsh requirements of materials used in various applications (*e.g.* transportation or armament). By combining this enhanced resistance of man-made assemblies with the hierarchical notions of biological materials, one can anticipate the creation of tantalizing responsive structures. When considering the latter, one must also keep in mind the necessity to direct the self-assembly processes and to have them happen at a fast rate. While biological materials assemble at a slow speed (often due to concurrent interactions), any sort of technologically relevant artificial assembly would require high-throughput procedures to warrant economical viability. While Nature can afford long times for assembly to sustain life, it is indeed desirable to operate at greater celerity as one aims towards the widespread implementation of smart materials (*i.e.* materials capable of adapting their environment). Indeed, such materials typically require long fabrication time and are consequently rather costly. Of concern also when designing next-generation mechanoresponsive polymers is the introduction of environmental cost in the equation of the design of said materials. Much like Nature chose to use limited raw materials, it would make sense in the long run to limit oneself to non-deleterious starting materials and/or materials that can be easily recycled/reused or degraded.

Whenever possible, the best approach to ensure the success of biomimetic strategies takes advantage of the synergistic combination of diverse expertise (*e.g.* biophysicists, to synthetic chemists, material scientists, computational chemists and bioengineers). It is only through the combination of these know-hows that we will succeed in developing not only passively smart materials (*i.e.* that will be pre-programmed to respond by means of set transitions) but also actively smart materials (*i.e.* whereby a feedback loop serves to modulate the response as proposed in Chapter 7 of this monograph).

1.2.2 Introduction of the Importance of Mechanotransduction Pathways for Living Organisms

The ability to sense force is of paramount importance for the survival and the favorable evolution of complex organisms comprised of many cells. The latter themselves are constantly experiencing a slew of mechanical actions: flow, elongation, pressure and pressure waves. The ability to change as a response to the nature of their surrounding environment is capital in the development and the subsistence of living organisms. This feedback is necessary for a variety of key biological functions such as proliferation, differentiation, motility or cell death.[11] It also plays a role in organ growth, bone adaptability and homeostasis (*viz.* the maintenance of parameters, *e.g.* pH, sugar, within a normal range for an organism). By analogy, for smart materials, one could anticipate systems that adapt to mechanical load or vibration, detect and report failure as well as initiate needed mending cycles. The implementation of feedback loops and graduated responses is a necessary condition for the future development of better smart materials. In living organisms, deficient feedback loops in mechanotransduction are responsible for numerous diseases such as cardiomyopathies, cancer metastasis or muscle degeneration. At the cellular level, mechanical deformation triggers diverse signaling pathways or induces changes in ionic concentrations, which in turn modulates the response appropriately (adjustment of the stiffness of the cell, its shape or even cross-talk with the extracellular matrix).

When one is thinking about the notion of mechanotransduction, the first things that probably come to mind are the senses of touch or hearing. Sensory cells are responsible for both. These cellular structures are hyper-specialized and have evolved to transduce mechanical inputs into given signals (*e.g.* flow of Ca^{2+} ions through transmembrane proteins to elicit an action potential in neurons). Sensory cells were logically amongst the first ones to be studied, as they offered a convenient model for mechanosensing. However, mechanotransduction goes far beyond the mere sensory pathways. It is important in the proper functioning of the vascular system and, in particular, in the role of the heart and its development.[12] It shapes the bones and their strength as a result of muscle contractions and external forces (*e.g.* gravity or hits). Think of the ability of a martial arts expert to strengthen his/her shins or forearms to endure greater impact and administer harsher blows, or even patients encouraged to walk to limit osteoporosis and encourage bone growth. This self-strengthening mechanism is highly desirable and has been the source of inspiration for exquisite studies in the field of synthetic mechanochemistry. For instance, Black Ramirez *et al.* have demonstrated the possibility to induce mechanochemical strengthening as a response to shear forces in a poly(butadiene) system containing *gem*-dibromocyclopropane along the chain and infused with the ditetrabutylammonium salt of sebacic acid.[13] Upon ring-opening of the three-membered propane cycle, one bromo-group thus formed can react with the dicarboxylate, thereby promoting the sought-after cross-linking reaction and subsequent reinforcement. Embryonal cell fate is also directed by external forces applied to the cell, which helps in the coordination of tissue growth. Also, said forces have been shown to influence stem-cell differentiation and have been proposed as a way to replace diseased tissues. Respiration cycles and the mechanical feedback loop associated with them contribute to the lung homeostasis.

Interestingly, in the body, all these circuits work in parallel, meaning that the overall response is the fruit of multiple mechanotransduction events. Wang *et al.* introduced the concept of gating in polymers as a means to control stress-response. The incorporation of cyclobutane served as a "gate" to control the activation of *gem*-dichlorodicyclopropanes. The single-chain force spectroscopy measurement indeed demonstrated the power of that gating mechanism for the design of next-generation materials. While this is rather limited in scope as compared to much more sophisticated biological systems, it is certainly a step in the right direction. Therefore, by combining mechanophores and/or specific architectures, one can readily envision creating systems that will get closer to emulating the intricate mechanotransduction pathways.

Mechanotransduction approaches can be broadly classified into those involved in sensory mechanisms (*i.e.* involving ion channels) and those mediated by more complex signaling pathways (*i.e.* entailing the coordination of a cascade of biochemical processes leading to the desired response). The former rely on the concerted rearrangement of vast supramolecular ensembles, such as lipid bilayers and pore-forming structures, while the latter are more often than not the consequence of protein unfolding and molecular conformational changes. The former are typically fast, while the latter often have latency times in excess of minutes. Interestingly, and as stated previously, both strategies involve non-covalent interactions. By reducing the number of bonds to break and form, Nature optimizes its energetic savings and limits the penalty associated with what would be a complex return to the pre-load state (equilibrium or metastable). Orr and coworkers have proposed the hypothesis that specialized cells employ mechanisms for mechanotransduction similar to the rest of the cells but obtain greater sensitivity by magnifying the strains experienced by the primary transducers.[14]

1.2.3 Adaptivity in Bones

Biological structures are not mere static systems. Instead, they are actively responding to the stimuli from their milieu,[15] constantly adjusting their architecture to adapt their properties to their surroundings. There are, for instance, repair mechanisms that are used to maintain the integrity of the system and improve its performance or simply to repair damage caused by external factors. These very principles constitute the basis of mechanobiology, which has the mantra "form follows function" as it assumes that an organism changes its structure to fit its role. Perhaps a good example of this mechanobiological approach is observed in bone development. In 1962, Chalmers and Ray carried out a classic experiment to investigate the influence, if any, of mechanical loading on the structure adopted by a bone during osteogenesis.[16] They used a cartilage model of a femur and transplanted it into the spleen – an area without mechanical loading. The cartilage developed into a femur, but did not have any of the refinements that are associated with a functioning femur (*e.g.* its characteristic shape), thereby illustrating the necessity of mechanical force in osteogenesis to fit function.

The bone is able to fine-tune its properties in response to changes in the mechanical environments thanks to the permanent remodeling of its constitutive material. This adaptation occurs on two different occasions: (i) during growth, or (ii) as a response to new mechanical requirements. Both can, however, be explained by the same remodeling process. In cancellous bone (*i.e.* the spongy portion of osseous tissue as opposed to cortical bone, its compact counterpart), a cavity is formed when osteoclasts (large bone cells that absorb bone tissue) resorb a "bone packet".[15,17,18] This cavity is then filled by osteoblasts to form a new bone matrix (Figure 1.1). Gradually, the matrix absorbs minerals, rapidly at first then slowly, increasing its stiffness. This bone remodeling process occurs throughout an animal's lifetime in response to mechanical strains exerted on the bone cells. Fratzl mentions that an adult skeleton contains between 1 and 2 million remodeling sites.[15]

In addition to the effect of mechanical loading during bone growth, a fascinating fact about bone tissue is the control that said mechanical forces have in the bone remodeling process. This role has been known as early as the late 19th century, with studies of cancellous bone structures showing that bone remodeling was a mechanically controlled process rather than a random replacement of old bone matrix. The Wolff–Roux law states that "bone is deposited wherever mechanically needed and resorbed whenever there is no mechanical use", in a "use it or lose it" system.[15] Frost further proposed that there is a feedback loop that regulates bone mass and architecture, functioning like a "mechanostat", where an increase in mechanical loading leads to bone deposition and *vice versa*.[19] This level of adaptability is evidently the Holy Grail of materials science where the system could

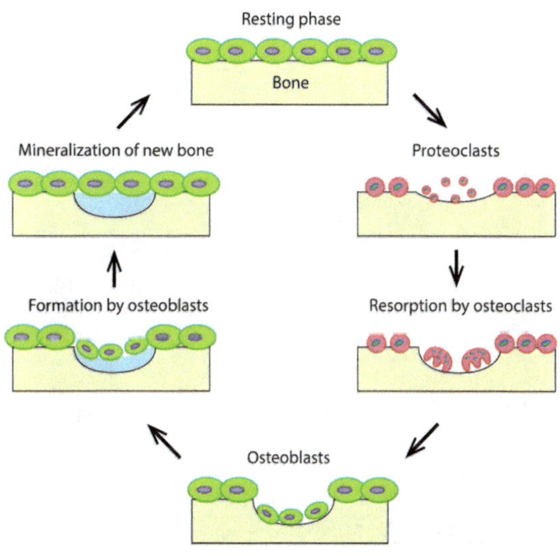

Figure 1.1 A schematic illustration of the bone remodeling model. Redrawn from ref. 18 with permission from The Royal Society of Chemistry.

readily stiffen or soften at will. In bones, increase in mechanical loading can be caused by a new intensive exercise activity (*e.g.* squash and tennis) while a loading decrease can be caused by inactivity. Frost estimates that increases in local strains of between 0.15 and 0.3% will lead to an increase in bone mass while resorption can be caused by a 0.01–0.03% decrease in local strains. As one considers the possibility to have switchable mechanomorphing materials, it is indeed interesting to note this asymmetry between growth and resorption, or as the case would be in man-made materials between stiffening and softening. Evidently, biological systems present the advantage of having a "constant" supply of building blocks brought on site through the vasculature, which is harder to implement due to the difficulty to implement such a system. It is, however, worth noting the remarkable work of Sottos, Moore, White and coworkers, on the fabrication of vascular structures for self-repair, whereby, for instance, dicyclopentadiene is polymerized *via* ring-opening metathesis polymerization to heal cracks.[20] Evidently, these vascular systems are based on damage and not simply mechanical force and also cannot currently help in the implementation of reversible softening and hardening through a desirable feedback loop. Nonetheless, these approaches could ultimately be used to circulate the necessary components to reversibly stiffen or soften as a response to mechanical force. Unlike bone growth, which requires a complex process, one can readily imagine a system whereby the kinetics of adaptation could be dictated by the kinetics of a given reaction (Figure 1.2).

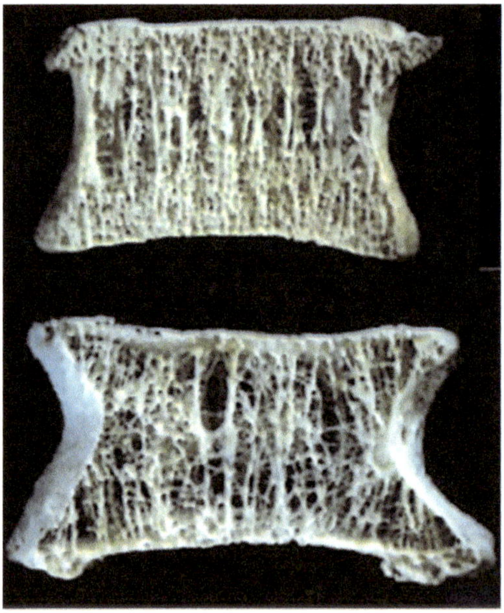

Figure 1.2 A comparison of bone mass content in normal (top) *vs.* reduced-gravity (bottom) conditions. Image credit: NASA.

In seeking to understand mechanical force impact on bone development, a number of experiments are often performed, probing factors such as frequency, magnitude and direction of the forces. An *in vivo* experiment is the avian ulna model (Figure 1.4a), where the ulna is only exposed to external loading. Results have shown that dynamic mechanical stimulus produces a bone forming response but a static stimulus does not. An interesting observation from this experiment is that bone loss can also be prevented by high frequencies (30 Hz) even when accompanied by sub-physiological stimuli. Interestingly, this cycling concept has not really been explored, except in materials science. Indeed, to the best of our knowledge, nothing has been done to reproduce what bones often experience, namely repeated impact at the limit of elasticity. By playing with frequencies and amplitude, one can anticipate some conditions under which mechanical activation would be enhanced. Similar observations could be made whereby deformations in the *a priori* elastic regime could upon cycling result in activation.

While proper load cycling has not been attempted, Craig and coworkers have demonstrated how soft spiropyrane-bearing polydimethylsiloxane systems could be used to detect damage by opening to the merocyanine form and reversibly closed to afford the "original" material (Figure 1.3).[21] Likewise, Imato *et al.* have used the reversibility in diarylbibenzofuranone-based mechanophores to achieve cycling in thermoplastic polyurethane.[22] Both of these examples take advantage of the reversible nature of the mechanophore linkages and repeated deformation is applied after a recovery step, whereby the broken link is regenerated.

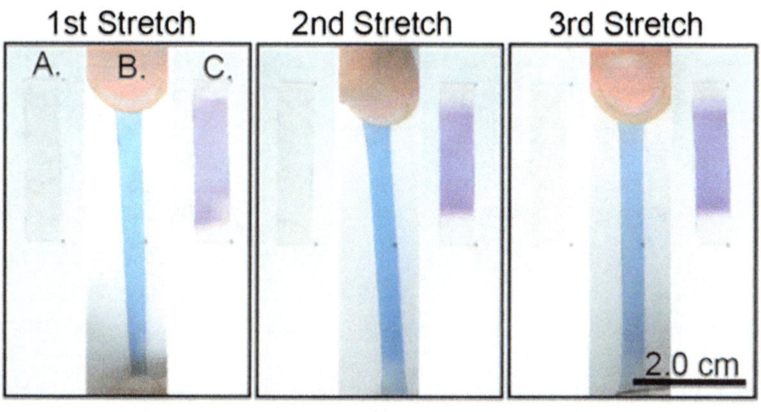

Figure 1.3 Covalent bond activation with full shape recovery. The original sample is clear and colorless under ambient conditions (A), but turns blue when stretched (B). When released, the material regains its initial shape (as indicated by the black reference spots), and the color switches to purple (C). Activation and shape recovery are repeatable over multiple cycles, as shown. Reprinted with permission from ref. 21. Copyright 2014 American Chemical Society.

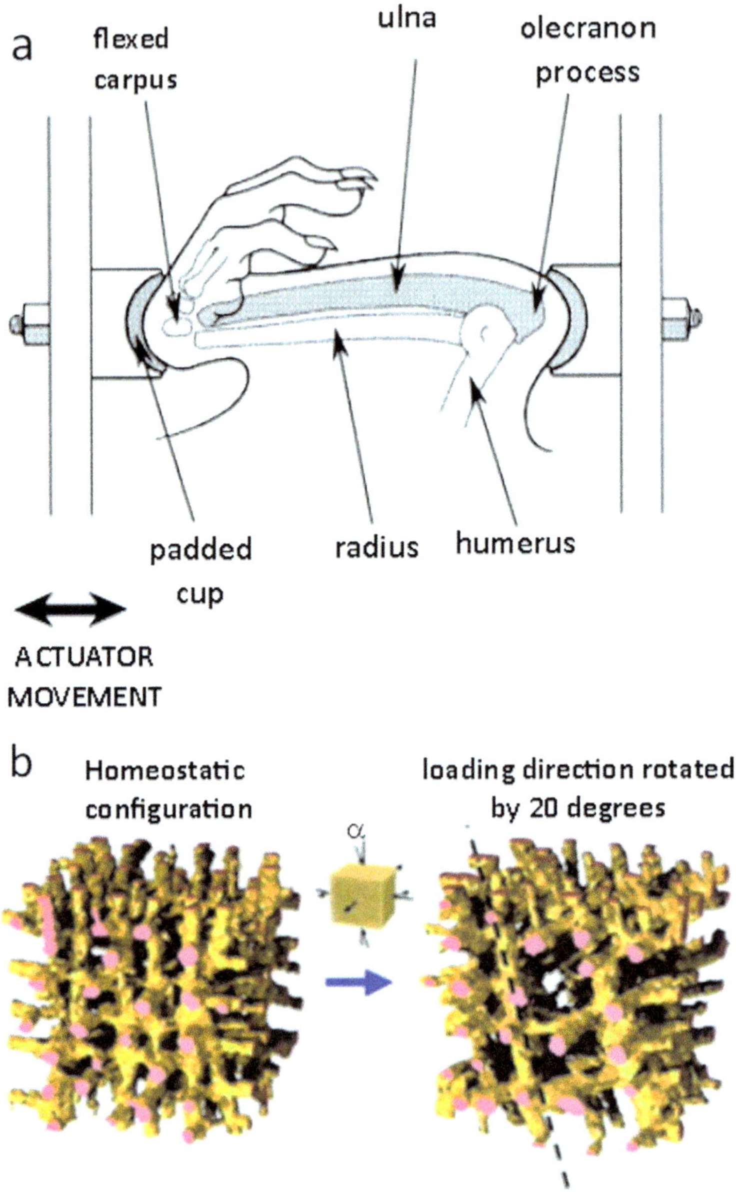

Figure 1.4 (a) Schematic representation of the non-invasive avian ulna model on a rat. Reprinted from *Bone*, 23 (4), J. R. Mosely, L. E. Lanyon, Strain rate as a controlling influence on adaptive modelling in response to dynamic loading of the ulna in growing male rats, 313–318, Copyright 1998 with permission from Elsevier.[24] (b) The effect of changing the external loading on a trabecula. Reprinted from *Journal of Biomechanics*, 38 (4), R. Ruimerman, P. Hilbers, B. van Rietbergen, R. Huiskes, A. theoretical framework for strain-related trabecular bone maintenance and adaptation, 931–941, Copyright 2005 with permission from Elsevier.[25]

Interestingly, Larsen *et al.* have used repeated activation to demonstrate the concept of flex-activated mechanophores, whereby multiple compressions after folding were used to achieve maximum activation. Here, again, this concept is different from that of oscillatory stress. Instead, the material is repeatedly damaged and plastically deformed. Also of interest though is the work of Beiermann *et al.*,[23] who looked into the alignment of chains (as a result of tensile stretch rate) as a factor impacting greatly the activation of mechanophores. They found that increasing stretch rates led to a greater accumulation of stress (*ergo*, a more pronounced response). One could anticipate that repeated application of stress in one direction may lead to a rearrangement of chains in one dimension and a more likely activation.

While most *in vivo* experiments like those mentioned above seek to understand the end results of the mechanical forces on bond cells, *in vitro* studies aim to relate the bone cells' biological response to the Wolff–Roux law and investigate the mechanotransduction pathway for such processes. They help elucidate how mechanical stimulus is sensed by the bone cells and transduced into cellular signals, which are then relayed to osteoclasts and osteoblasts. In bone cells, small channels called canaliculi are responsible for the supply of nutrients to the osteocytes. Loading of bones will cause interstitial liquid to be squeezed through the canaliculi, leading to stimulation of the osteocytes by both the fluid shear stress and electromagnetically due to the ionic composition of the tissue fluid. Studying these mechanotransduction processes through has also shed more light on the conditions that facilitate osteoporosis. For example, in response to pulsating fluid flow, osteocytes have been observed to produce higher levels of PGE_2, a signaling molecule, which promotes anabolic actions leading to bone formation. PGE_2 response to strain has also been shown to be enhanced by estrogen. This observation has been used to explain why there is a higher rate of osteoporosis in women during menopause. Response to assemblies to fluid motion is indeed a very interesting concept that has been investigated by Holme and coworkers[26] and then further revisited by the Zumbuehl group (Figure 1.5).[27] One idea is to create shear-sensitive liposome assemblies that will deliver their therapeutic payload as a function of shear stress in a liquid environment. For instance, one may consider such delivery systems for the treatment of vasodilation or clot lysis.

Another interesting feature of bone cells is the unique healing process, which occurs in response to fracture. The bone system is exceptional when compared to other tissues in that it is able to completely restore the original bone form and functionality instead of leaving scar tissue. This regeneration occurs in steps similar to the skeletal development process. On the extreme side, living organisms such as salamanders can actually regrow their limbs completely. While complete regrowth is not possible (yet?!) in humans, the bone-healing process still manages to replace/repair damage. In addition to biological influences, mechanical stimuli play an important role in the healing process. In a primary fracture, the fragments are first immobilized and the remodeling occurs through the creation of new bone cells to connect the two fragments. While the healing process is considered ideal, it is a very delicate process. The fragments are required to be free of any mechanical loading and

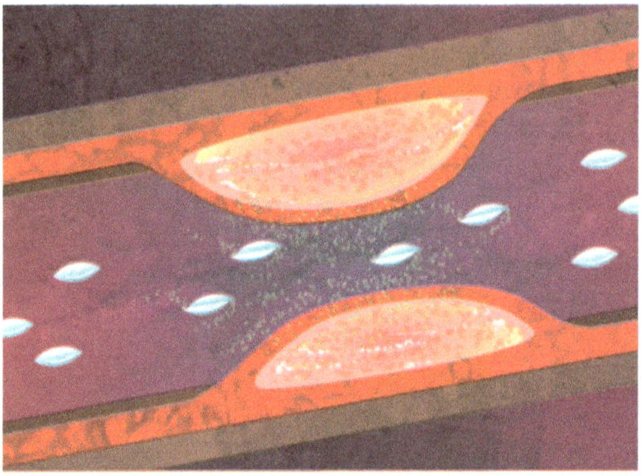

Figure 1.5 Nanocontainers liberate an active drug at the critically stenosed vessel because of the significantly increased wall shear stress. Reprinted with permission from T. Saxer and A. Zumbuehl, The use of shear stress for targeted drug delivery, *Cardiovascular Research*, 2013, **99**(2), 328–333, by permission of Oxford University Press.[27]

the bone is very weak until complete healing. However, since all the treatment procedures require some degree of motion, primary fracture healing is very rare. Secondary fracture healing, on the other hand, is less delicate. During bone fracture, the blood supply is disrupted, leading to the cell death. The inflammation that occurs on the affected area is instrumental in cleaning off the dead materials, restoration of the blood vessels and to congregate mesenchymal stem cells (MSCs). As will be discussed later in this chapter, mechanical forces are instrumental in the differentiation of these MSCs into cartilage, fibro-cartilage and bone cells. A hard bone shell forms around the fracture site, which is filled with cartilage. The soft tissue mineralizes and is finally replaced by bone, before the shell is removed by resorption. While the biological machinery involving multiple step recovery (*e.g.* inflammation, clean up, resorption) is hard to replicate with materials, it is still interesting to point out some of the principles behind healing as one considers ways to emulate Nature. For instance, the principle of using a scaffold for rapid repair followed by a resorption of said scaffold is an interesting concept that has not been exploited. It is, however, worth mentioning the work of White *et al.*, whereby a gelation step facilitates the recovery of large damage volumes, which in spite of the lack of resorption uses the principle of scaffolding (Figure 1.6).[28]

Studies of bone healing in Nature influence the design of self-healing materials. Remarkably, the effect of mechanical loading in bone cells can be used as an inspiration to develop materials whose function determines their structure or those capable of self-healing in response to damage by mechanical forces.[29] The latter has seen a number of materials developed, with potential use of self-healing materials in electronics, coatings and transportation industries. These systems are usually designed using three approaches: capsule-based, vascular (*vide supra*) and intrinsic methods. In the capsule-based method, a healing agent is stored in capsules. Upon material fracture, the capsules

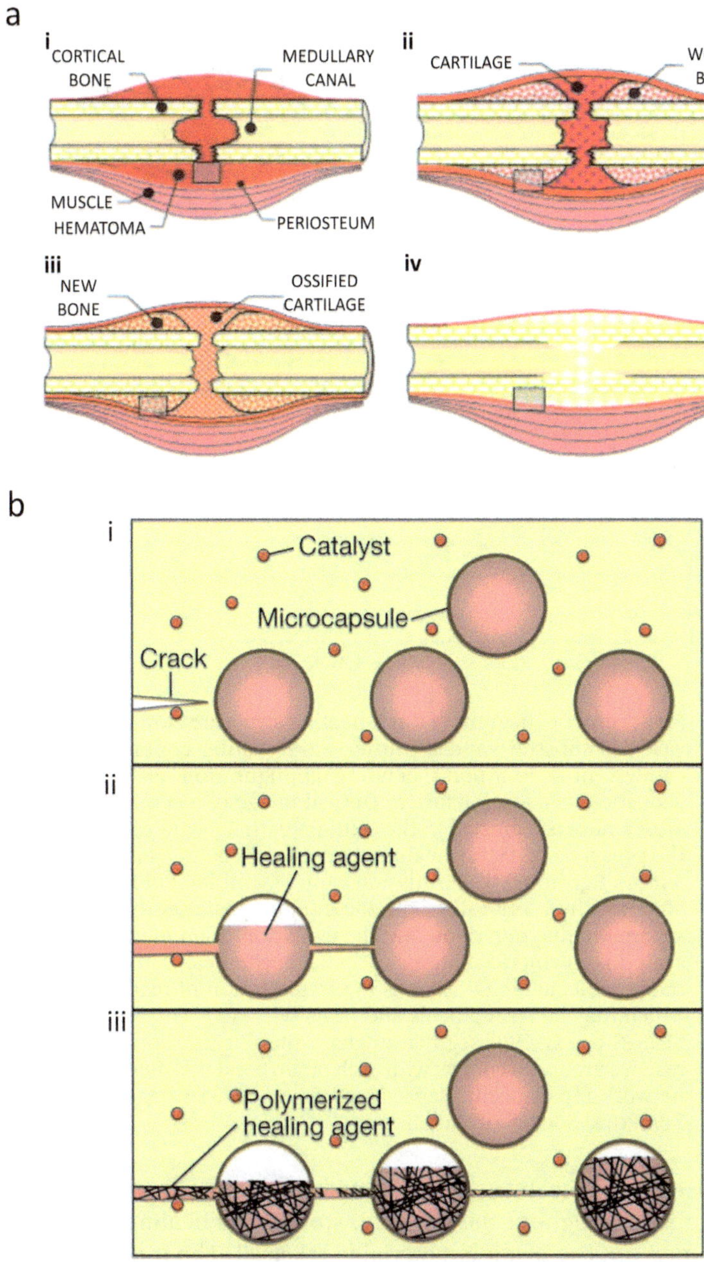

Figure 1.6 (a) The different phases observed during fracture healing. Reprinted from *Progress in Materials Science*, 52(8), P. Fratzl and R. Weinkamer, Nature's hierarchical materials, 1263–1334, Copyright 2007 with permission from Elsevier.[15] (b) A self-healing epoxy matrix. Reprinted by permission from Macmillan Publishers Ltd: *Nature* (ref. 20). Copyright 2001.

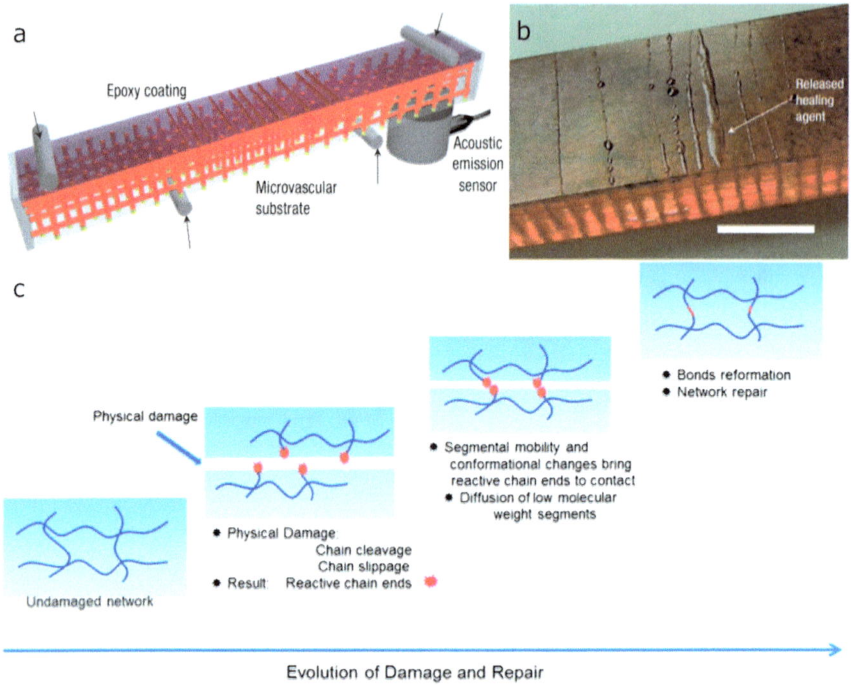

Figure 1.7 (a) Schematic diagram of a self-healing structure composed of a microvascular substrate and a brittle epoxy coating containing embedded catalyst in a four-point bending configuration monitored with an acoustic-emission sensor. (b) Optical image of a self-healing structure after cracks are formed in the coating (with 2.5 wt% catalyst), revealing the presence of excess healing fluid on the coating surface (scale bar = 5 mm). Reprinted by permission from Macmillan Publisher Ltd: *Nature Materials* (ref. 30). Copyright 2007. (c) Intrinsic self-healing system. A schematic diagram depicting the evolution of an ideal damage–repair cycle in polymeric materials. Upon mechanical damage, cleavage of macromolecular chains leads to the formation of reactive end groups, which may be free radicals and/or –C=C–, –COOH, –NH2, –OH, –Si–O, SH/S–S or –C=O. If segmental chain mobility and/or diffusion brings reactive groups in contact with each other, bond reformation and physical network repair occur. Reproduced from ref. 31 with permission from The Royal Society of Chemistry.

rupture, releasing the healing agent. The vascular approach employs a matrix of hollow tubes throughout the material, storing the healing agent (Figure 1.7). Upon fracture, the healing material is released. The intrinsic self-healing polymeric systems often contain materials with a latent functionality that can trigger the healing process in the event of damage.

One of the earliest examples of a capsule-based self-healing system was demonstrated by White *et al.* in 2001 with an epoxy matrix.[20] The matrix was loaded with urea-formaldehyde shells encapsulating dicyclopentadiene (DCPD) as the healing agent. A small amount of a Grubbs catalyst was dispersed in the matrix and, upon cracking of the matrix, the microcapsules were observed to rupture, releasing the healing agent. Upon contact with the Grubbs catalyst, the DCPD polymerized, rebonding the cracked matrix. Since then, many other groups have exploited the principle of capsules for damage reporting.[32–35]

1.3 Sensing

1.3.1 *Via* Protein Unfolding

For decades, studies in various biological systems have shown how mechanical forces on the small scale lead to responses from simple microscale levels in cells to large-scale ones that can lead to even larger consequences like changes in transcription and protein levels. The mechanical stimuli responsible for these changes includes hydrodynamic forces, shears and stresses. Early work by Klibanov and Ishimori showed that often simple enzyme conformational changes brought upon by mechanical forces can lead to changes in enzyme activity.[36] Klibanov *et al.* covalently attached chymotrypsin and trypsin to nylon fibers and observed that when stretched, the enzymes lost their ATPase activity. Later, Ishinori also observed a decrease in activity on glucose oxidase attached to polyvinyl chloride membranes. From these observations, they hypothesized that enzyme deformation due to the stretching were responsible for the loss in activity. The influence of mechanical force on enzyme is actually no stranger to the field of polymer science as this was the topic of the thesis of Tirrell in 1977.[37]

This reliance on mechanical perturbations by various enzymes has also been observed in proteins that possess cryptic sites or sites that are hidden when in the protein's natural state. The presence of mechanical force leads to unfolding of the proteins exposing the hidden reactive sites that are responsible for initiating signaling pathways. One such example is the cytoskeletal protein, Talin, active in the cell adhesion process.[36] Talin is responsible for linking integrins to the cell's cytoskeleton by binding to vinculin, another cytoskeletal protein. In its resting state, the binding locations are inaccessible to the vinculin but when stretched by tensional forces, the sites become accessible leading to the vinculin binding and force transfer to the extracellular matrix (ECM). Similarly, the von Willebrand Factor (vWF), a protein instrumental in the blood clotting process also uses the cryptic site approach. In the case of vascular damage, the hydrodynamic forces from the blood lead to a conformational change from the vWF's coiled state to an unfolded state, which exposes the binding sites for platelet receptors, which play an important role in the blood clotting process. Fibronectin (FN), a protein instrumental in extracellular matrix (ECM) binding to the cells, also exhibits these cryptic sites as shown in Figure 1.8.[38] In FN, the spatial positioning of its loops (RGD and PHSRN) controls the cell binding. Small forces can unfold FN, increasing the distance between the two loops, and exposing its binding sites in its type III domains. In this stretched state, $\alpha5\beta3$ integrin binding is favored while the unstretched state allows for $\alpha v\beta3$ binding. In a simple way, Nature uses changes in protein conformation to control access to binding sites, thereby resulting in the ability to regulate complex biochemical process using low energy. These cryptic-site-based systems show how small mechanical forces can elicit large-scale responses in the body.

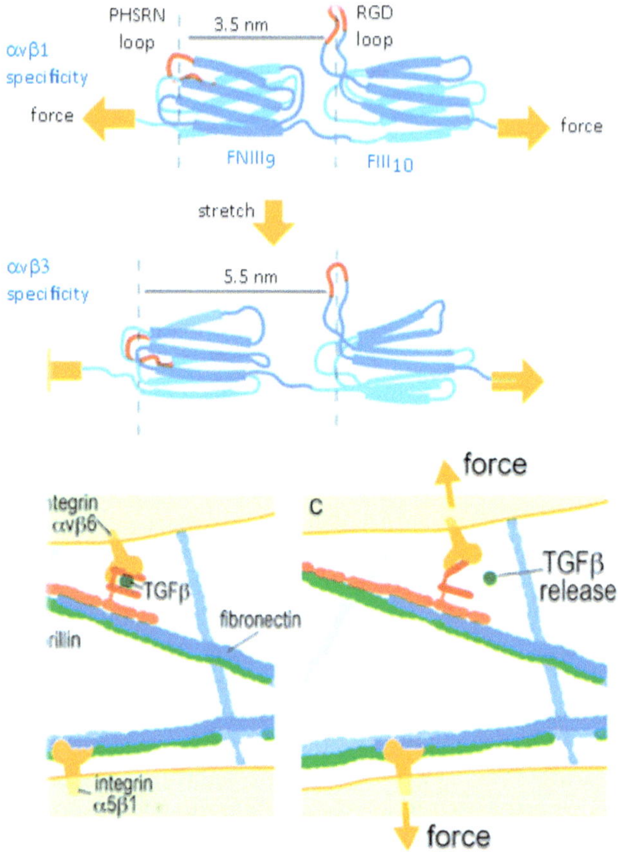

Figure 1.8 Interaction between cell binding and mechanics as illustrated by the spatial positioning of RGD and PHSRN loops in fibronectin. Forces in the tens of pN are sufficient to modulate integrin specificity by altering the distance between these two loops. Reproduced from ref. 38 with permission from The Royal Society of Chemistry.

Work done by Discher *et al.* in their investigation of red blood cell (RBC) protein, β-spectrin, also showed how mechanical forces can influence specific binding of a protein to cysteine residues.[39] In RBCs, blood flow induces stresses that can lead to deformation. The membrane of RBC is reinforced by a cytoskeleton constituted in part of spectrin chains connected to F-actin through its cysteine residues (Figure 1.9).[40] F-actin is an essential protein instrumental in the deformation process. Spectrin's α and β chains contain 20 and 15 cysteines, respectively. In the experiment, RBCs were reversibly lysed and filled with a reactive cysteine-modified fluorophore before being reformed and sealed ("ghost RBCs"). They were then exposed to a range of conditions at varying temperatures, while some were sheared using a microfluidic device. They were then relysed and the results showed that the sheared RBCs had about 66% more fluorescent dye compared to the static ones, suggesting that the folding hides specific domains.

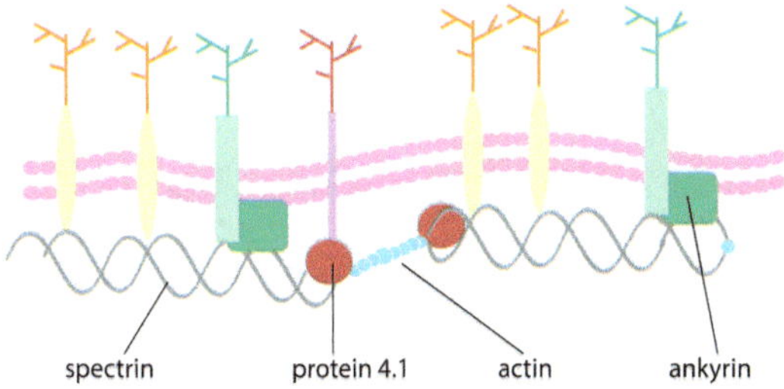

Figure 1.9 The spectrin membrane skeleton. The classical spectrin membrane skeleton of the erythrocyte (RBC) is composed of linear spectrin oligomers cross-linked by short actin filaments to form a two-dimensional meshwork. This meshwork is coupled to the cytoplasmic surface of the plasma membrane through a variety of interactions: (1) direct interactions of membrane proteins with the β-subunit of spectrin; (2) binding to the membrane skeleton protein ankyrin; and (3) binding to the membrane skeleton protein 4.1, a protein which also serves to mediate spectrin–actin interactions.[40]

The biomimetic inspiration from these simple cryptic site systems has been adopted in the rising field of soft-mechanochemistry, which is carving out a new approach in polymer and materials science design. The idea of mechanochemistry, the process of transforming a mechanical force into a (bio)chemical response has been around for a long time. However, most systems often require harsh conditions in the form of "high-energy forces and high energy input" and are mostly irreversible. Therefore, the possibility of using small mechanical forces to elicit large changes has the ability to significantly increase the application of polymer science in both biological and non-biological systems. Impressive work in soft-mechanochemistry has been done by Jierry and coworkers at the University of Strasbourg, whereby they have demonstrated polymer-based systems that utilize cryptic sites to reduce the energy demand.[36] Using a surface-functionalized elastic layer of polydimethylsilane (PDMS), they grew polyethylene glycol (PEG) brushes on the surface before attaching biotin, a ligand for the protein streptavidin. In the unstretched state, the PEG brushes completely covered the biotin on the surface. When a streptavidin solution was added to the surface, no binding was observed, but, when stretched, the streptavidin bound to the biotin, with a linear increase in binding with stretching observed. This binding was reversible, with the strong lateral pressure from the PEG brushes breaking the strong non-covalent bond between the streptavidin and the biotin. When the biotin was replaced with RGD, a cell adhesion protein, the setup showed a reversible binding to osteoprogenitor cells.

These cryptic site-based systems have also been successfully used to design mechanoresponsive catalytic systems by the same group. In one such system, they developed a catalytically active 3-D network with partial reversibility by using a maleimide-modified version of the enzyme B-galactosidase (Figure 1.10a).[41] The network consisted of polyelectrolyte multi-layer films (PEMs) of thiopyridyl-modified poly-L-lysine (PLL-S-TP) and hyaluronic acid (HA)

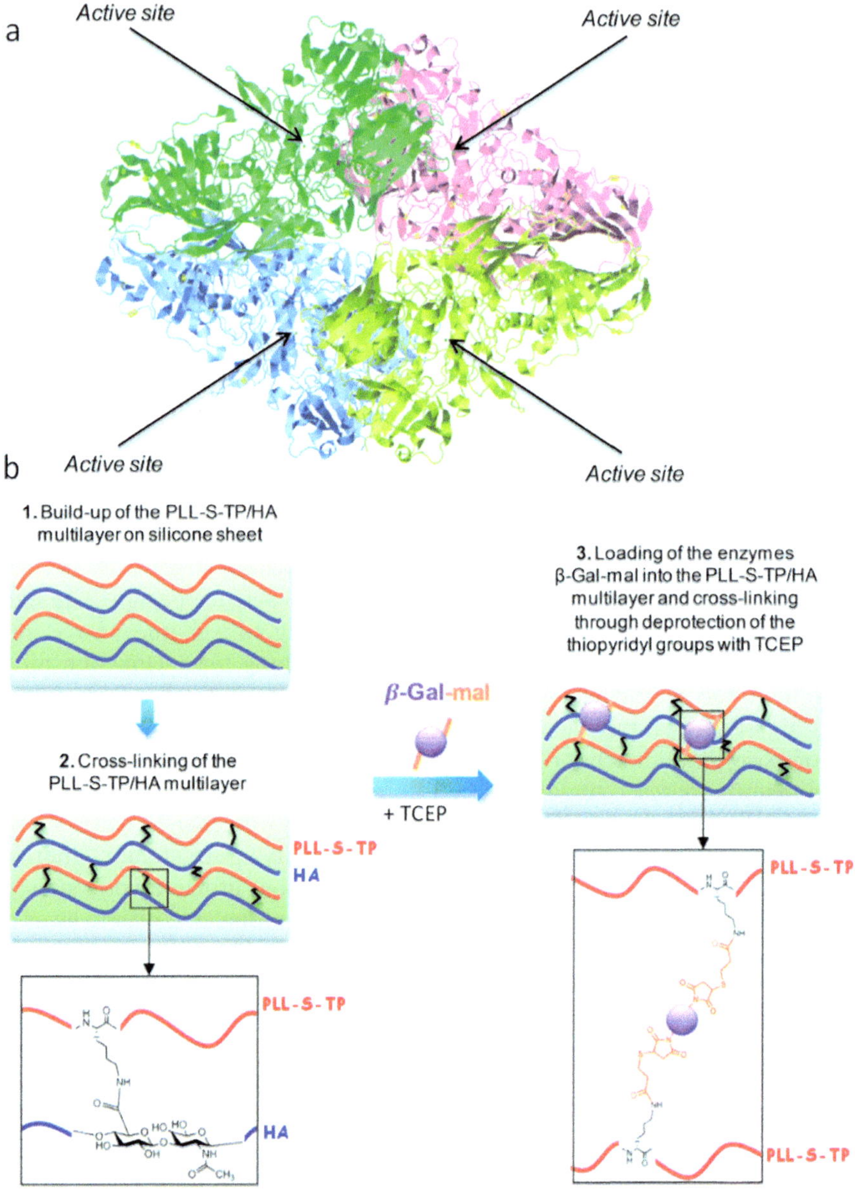

Figure 1.10 (a) The four different active sites of the β-galactosidase enzyme. Image originally taken from the PDP (code 1BLG). (b) Schematic representation of the pathway taken in the development of the enzyme-based mechanosensitive catalytic system. Reproduced from ref. 41 with permission from The Royal Society of Chemistry.

deposited on PDMS. After initial carbodiimide cross-linking in the PEMs, the modified enzyme was then introduced, and diffused into the layers before permanently embedding into the layers through the thiol-maleimide click reactions with deprotected thiopyridyl moieties as shown in Figure 1.10b. Stretching the PEM films resulted in a catalytic activity decrease of up to 40% (with fluorescein di(β-D-galactopyranoside) as the substrate). They hypothesized that the stretch-induced enzyme conformational change led to the decrease in activity.

Perhaps the earliest influence of mechanical forces in human growth is evident in the heart, the first functional organ to develop in an embryo. Cardiogenesis relies on the stresses exerted on the heart for growth and cell differentiation. In trying to understand cardiac tissue development, Discher *et al.* postulated a model centered on cell-matrix interaction in which the proliferation of fibroblasts is limited by the stiffness of their environment, which in turn correlates with their collagen matrix density.[12] Collagen production is mainly done by fibroblasts; therefore, collagen mRNA is expected to be proportional to the fibroblast composition. The fibroblast composition is determined by the stiffness of the environment, itself controlled by the ECM. The overall point of the model is that turnover (*i.e.* expression and lysis) of key structural proteins is mechanoregulated. Cells transduce stresses/strains or any forces that are exerted on them into a change in protein level, which is reflected at the transcription level (hence the link to RNA). A model which supports the idea that forces affect protein synthesis/degradation and transcription means that a "use it or lose it" mechanism likely controls key circuits.

The approach based on Discher's model seems to suit the observations that have been made during growth from an embryonic heart to a full adult. The density and stiffness of the extracellular matrix (ECM) regulates the function and structure of the neonatal cardiomyocytes (cardiac muscle cells). In a seemingly self-reinforcing model for heart development (Figure 1.11), the stresses exerted through neighboring cells is propagated through the ECM and cell–cell adhesions onto cardiac fibroblasts (connective tissue producing collagen). In response to the strain, the fibroblasts divide and produce more ECM, which in turn cause an increase in production of microfibril proteins. The resulting reorganization of the microfibril produced exerts a strain on the fibroblasts, restarting the cycle. This somewhat simplified model of fibroblast and cardiomycyte balance during cardiogenesis can provide ideas on the design of "self-aware" systems, which use the presence of mechanical forces around them to control their growth or perform their functions.

Research in muscular dystrophies and other conditions linked to skeletal muscle cells has also helped elucidate the importance of mechanotransduction and the complexity of the pathways that allow for signal transmission from the extracellular matrix to the nucleus to elicit responses at a transcription level.[11] The transmission of extracellular forces into signals that reach the nucleus is initiated in the extracellular matrix (Figure 1.12), which as mentioned earlier consist of tissue-specific proteins such as collagen, laminin and fibronectin. The ECM is linked to the cytoskeleton (actin) through adhesion complexes that are present at the cell surface. In skeletal muscle, this adhesion complex consists of dystrophin and a dystrophin-associated protein complex

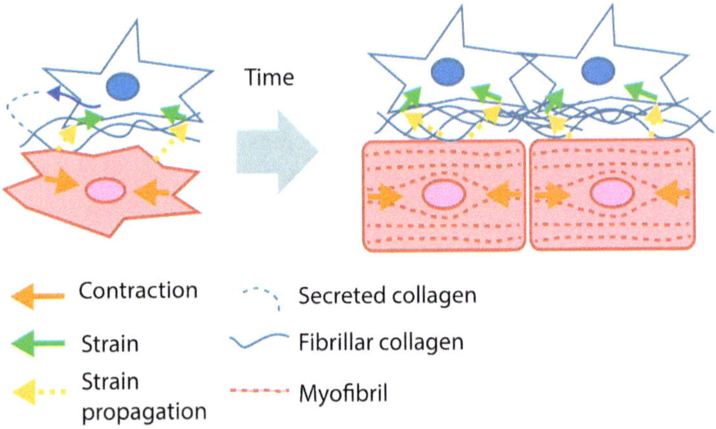

Figure 1.11 A schematic illustration of the relationship between cardiomyocytes and cardiac fibroblasts. This relationship is supported by mechanosensitive pathways that occur through the extracellular matrix (ECM). Reprinted from *Current Biology*, 24 (10), S. Majkut *et al.*, Stress sensitivity and mechanotransduction during heart development, R495–R501. Copyright 2014 with permission from Elsevier.

(DAPC). Forces transmitted through the cytoskeletal network are then transferred to the nucleus through coupling to nesprins and other proteins on the outer nuclear membrane. Nesprins interact with inner nuclear membrane proteins (such as SUN 1 and SUN 2), which in turn interact with lamins and other nuclear envelope proteins such as emerin. Lamins complete the force transmission pathway by forming stable structures which are able to bind to DNA. A number of studies targeting different components of this mechanotransduction pathway have shown that changes in any of these sites could result in altered cellular functions. These changes can be brought about by factors such as mutations, and changes in cellular environments or structure. From a materials science perspective, it is interesting to see the hierarchy and the level of modularity found in living organisms. Quite manifestly, it is thus far nearly impossible to emulate this level of intricacy. Nevertheless, as we aim to fabricate ever more complicated systems, it is quite intriguing to notice that parallel pathways involving multiple components are involved in graduated responses.

The genetic disorder Duchenne muscular dystrophy is an example of the effect of changes in one component of the pathway. Mutations in the gene responsible for encoding dystrophin production lead to disruption in the dystrophin-associated protein complex's ability to transmit force signals between cytoskeleton and the ECM, eventually leading to progressive muscle degeneration and other consequences in cell function and viability. For example, work by Loufrani *et al.* shows that muscle-fibers with a dystrophin deficiency lead to decreased vascular density in cardiac muscle.[42] Defects and mutations in the nuclear envelope proteins such as nesprins and emerin have also been shown to result in muscular dystrophies. In general, disorders that are caused by disruptions in the mechanotransduction pathways are often caused by changes in

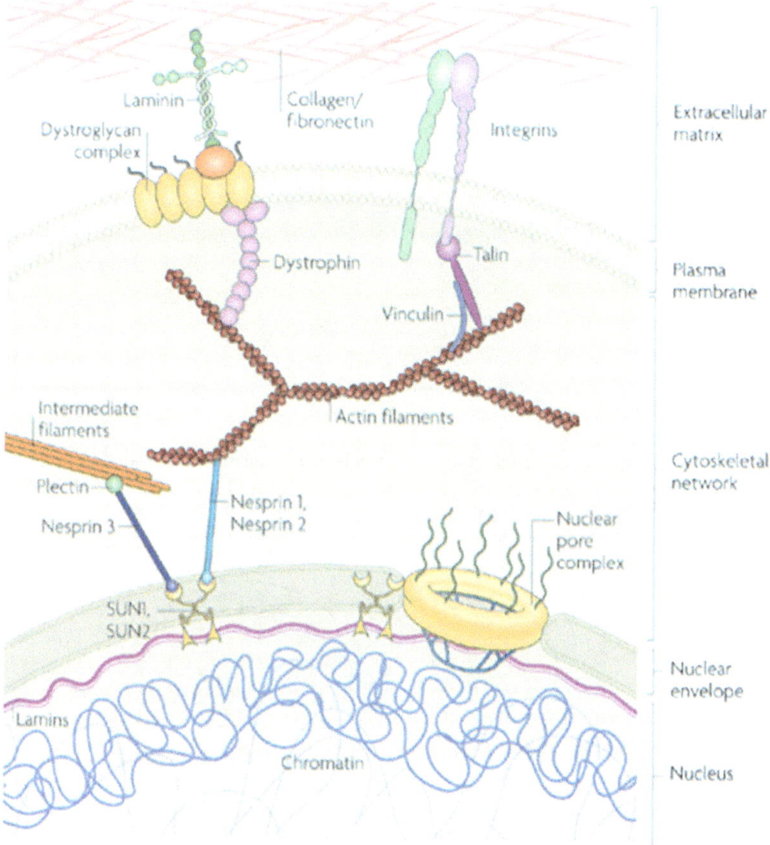

Figure 1.12 Schematic illustration of the pathway used for force transmission from the extracellular matrix to the nucleus in skeletal muscle tissue. Reprinted by permission from Macmillan Publishers Ltd: *Nature Reviews Molecular Cell Biology* (ref. 11). Copyright 2009.

three components that directly affect the pathways: the extracellular environment, cell structure and organization, and defects in the sensing and signaling (Figure 1.13).

As Figure 1.13 shows, the number of possible factors that can lead to defective mechanotransduction pathways cannot be overstated, and therefore the study of the effect of mechanotransduction in physiology is clearly important. Understanding the role of mechanotransduction in systems can lead to an increase in the role polymer scientists can play in the design of possible remedies.

1.3.2 *Via* Ion Channel Opening

Mechanosensory hair cells convert mechanical energy from sound waves into electrochemical signals that can be further transferred into the nervous system. Mechanically gated ion channels in the mechanotransduction

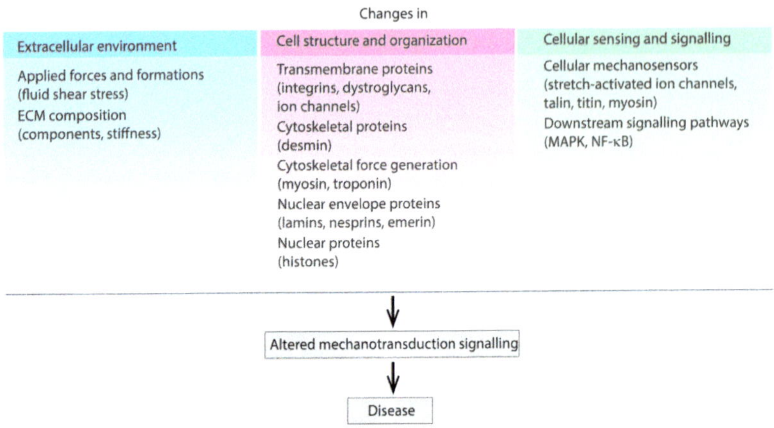

Figure 1.13 Common characteristics of mechanotransduction disorders. Reprinted by permission from Macmillan Publishers Ltd: *Nature Reviews Molecular Cell Biology* (ref. 11). Copyright 2009.

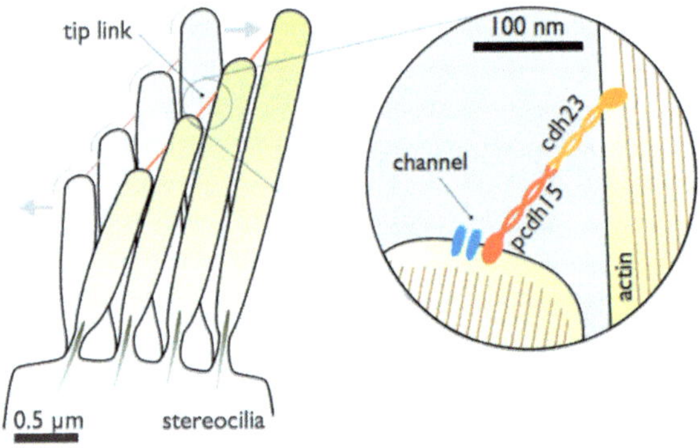

Figure 1.14 Stereocilia pivot when they are mechanically stimulated by sound. The tips of the cilia are linked by protein tethers made of cadherin 22 and protocadherin 15. This tip link is anchored within the membrane to an ion channel. A deflection of the stereocilia opens and closes the ion channels, which results in changes of ion flux across the membrane. A sufficiently strong deflection will eventually depolarize the cell and lead to an electrical potential that can reach the auditory nerve. Reproduced from ref. 46 with permission from The Royal Society of Chemistry.

pathway are located at the tip of the stereocilia (apical modifications of the cells), giving the hair cells a specialized sensitivity that is not observed in other cell types in response to mechanical forces (Figure 1.14).[43,44] The stereocilia are connected to the mechanotransduction channels by tip links, extracellular filaments which are able to transmit information on the differential deflection of the stereociliary tips to the transduction channels. The tip links' ability to sense direction of the hair stereociliary tips is understood to be instrumental in the state of the ion channels. Deflection towards taller stereocilia activates the opening of the ion channels and, consequently, a cation influx while deflection towards the shorter stereocilia leads to gate closure.[45] However, these characteristics are only a small part of the mammalian hair cell mechanotransduction puzzle. The transduction pathway is complex and consists of many components. While several of these components are known, there is not enough information on their roles in the mechanotransduction pathway. Recent models that have postulated that the complexity of the process is necessary for frequency discrimination of the mechanotransduction pathways have been identified.

Knowing more about hair cells and the transduction pathway will help with understanding and coming up with solutions and remedies for malfunctions in the mammalian cochlea. Of greater interest, however, is the potential of systems that can be developed from this knowledge. Developing systems that have the discriminative mechanosensitivity possessed by hair cells will open up exciting applications in both sensing systems and noise-cancelling systems. One such system was recently reported by Asadnia *et al.*, who developed novel "miniature all-polymer flow sensors that closely mimic the intricate morphology of the mechanosensory ciliary bundles in biological hair cells" (Figure 1.15).[47] Using the flexible polydimethylsilane (PDMS), they fabricated rows of pillars to mimic hair bundles in fish. Mimicking the geometry of the hair bundles imparted flow direction sensing on the bundles. The PDMS pillars were then connected to piezoelectric polvinylidine fluoride (PVDF), which acted as tip links. A hyaluronic acid-based hydrogel was then used to cover the pillars to enhance their sensitivity. This simple, all-polymer system showed sensitivity to water flow with a very low threshold detection of 8 $\mu m\ s^{-1}$.

In addition to our audial ability, the sense of touch is another fascinating example of human senses for which mechanotransduction plays a starring role. While conventional touch is essential for survival, allowing one to observe and respond to mechanical perturbations in their surroundings,[48] discriminative touch allows us to accomplish uniquely human operations, from playing musical instruments to tweeting about successes in the lab. A number of mechanosensitive neurons innervate the skin,[49] with each of the sensations often observed by the skin being mediated by different receptors. For example, light touch is mediated by Aβ afferents with a lower mechanical threshold. Painful touch, on the other hand, is mediated by nociceptors that have a higher mechanical threshold.

While there is still a large gap in knowledge about the mechanism of the transduction pathways, it is agreed that mechanical stimuli play an active role in directly activating these transduction channels. This mechanical gating can be activated in a number of ways, such as the use of forces in the membrane bilayer, or through the extracellular membrane (Figure 1.16).

There are four main classes of mechanoreceptors and their classifications are determined by factors such as morphology, sensitivity to dynamic and static events, mechanical threshold, firing pattern and the mechanical stimulus to which they respond.[50,51] These classifications are slow adapting-type I

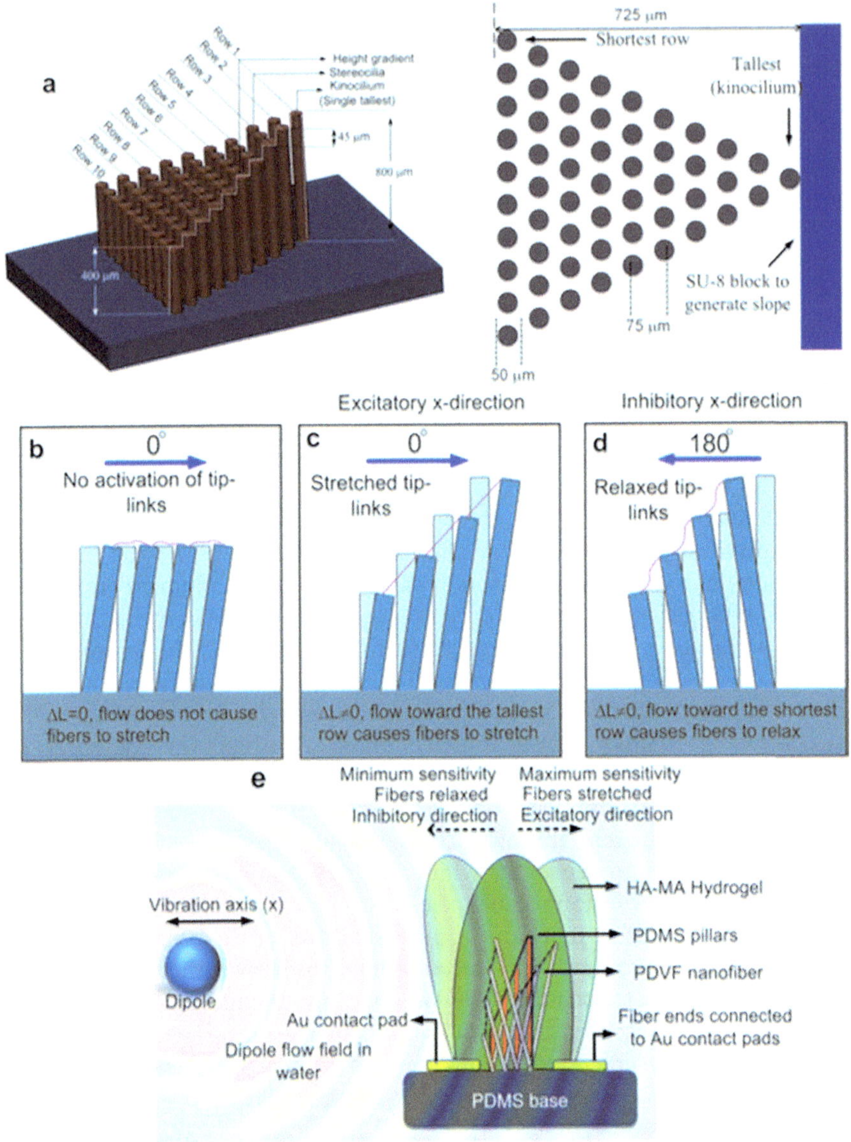

Figure 1.15 All-polymer flow sensors inspired by hair cells. (a) Angled side-view and top-view of the graded height and geometrical dimensions of PDMS pillars within a MEMS bundle. (b–d) Schematics illustrating the necessity of height gradation in the design of a pillar bundle and the directional dependence of the output, for different directions of flow along the bundle's axis. (e) A schematic showing the flow sensing mechanism in the presence of an oscillating dipole. Reprinted by permission from Macmillan Publishers Ltd: *Scientific Reports* (ref. 47). Copyright 2016.

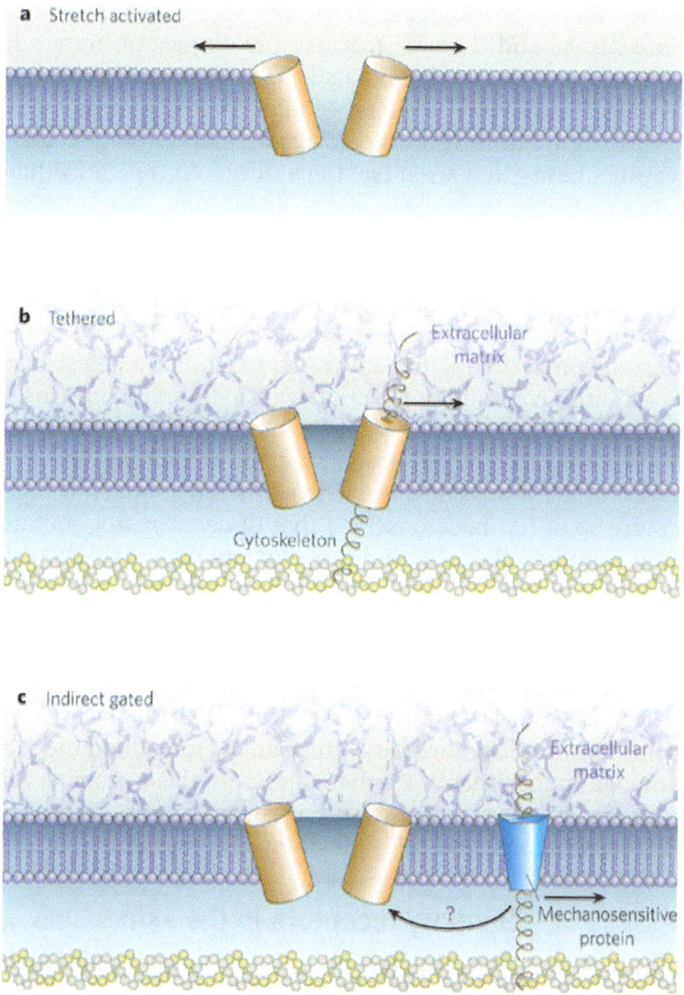

Figure 1.16 An illustration of the gating models in mechanotransduction. The ion channels can be opened (a) by forces from the membrane bilayer, (b) by strains in the cytoskeleton or extracellular matrix or (c) through the use of a signaling intermediate. Reprinted by permission from Macmillan Publishers Ltd: *Nature* (ref. 49). Copyright 2007.

(SA I), slow adapting-type II (SA II), rapidly adapting-type I (RA I) and rapidly adapting-type II (RA II). SA I receptors are often represented by Merkel cells, which are critical for detection of texture. SA II afferents consist of Ruffin endings and perceive the stretch of the skin between fingernails to determine an object's shape and for motion detection. FA I receptors consist of Meissner's corpuscles, which are responsible for low-frequency vibrations, slip and motion detection. FA II receptors, represented by Pacinian corpuscles, show high-frequency sensitivity. These afferents can also be classified based on their electrophysiological properties such as action-potential propagation into Aβ, Aδ and C-fibers. Just like the studies of human hair cells, the understanding of the mechanotransduction pathways in touch sensitivity still lags behind, on a molecular and cellular level. However, recent discoveries such as the identification of molecular markers for several types of touch receptors have given scientists the ability to probe individual sensory neurons. Interestingly, to the best of our knowledge, this approach relying on transmembrane gating proteins has not been pursued as an avenue for mechanochemistry in polymers (Figure 1.17).

Piezo proteins, which are a group of mechanically activated ion channels, are a class of recently discovered proteins that have provided more answers to the transduction pathways in mammalian touch.[48] They exist as tetramers, with each tetramer consisting of over 30 transmembrane proteins. They can be activated by a variety of stimuli such as touch, suction and shear stress, which points to the fact that they are activated by membrane deformation. Isolated Piezo1 proteins have been observed to form ion-conducting pores in lipid bilayers. More so, it has been observed that disease-causing mutations slow these gating channels. Piezo2 proteins, which are part of this protein family, are known to be responsible for mechanotransduction in most receptors. High concentration of the Piezo2 protein transcripts in dorsal root ganglion (DRG) neurons suggests that the gene is responsible for encoding transduction channels in touch receptors. Loss of Piezo2 proteins has been shown to lead to complete loss of gentle-touch responsiveness. The discovery of these Piezo proteins further helps to illuminate the role of mechanical forces and their transfer in the mammalian touch process. Further studies on these proteins will help us to better understand the mechanotransduction pathway in mammalian touch, and provide inspiration in the design of other systems in mechanochemistry.

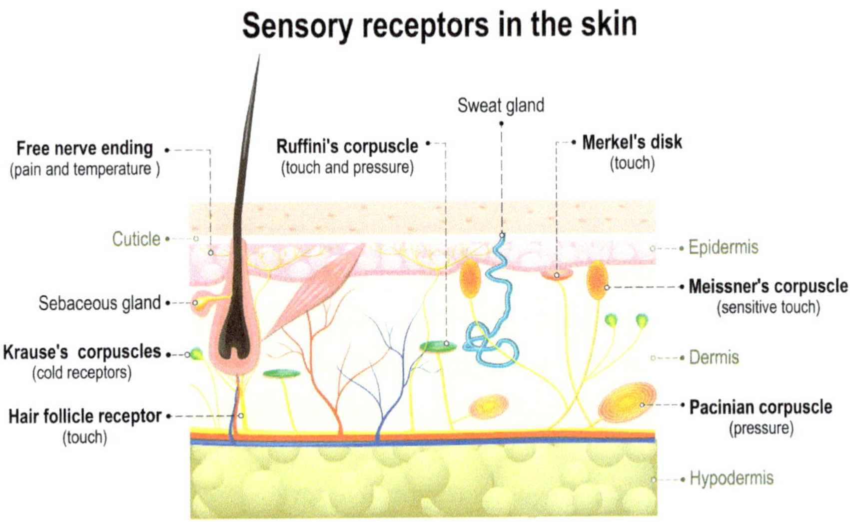

Figure 1.17 Illustration of the types of receptors observed in the hairless skin of the fingertip. © Shutterstock.

Just like hair-cell sensitivity, the ability to mimic touch sensitivity has a vast range of applications, from prosthetics to sensors. Consequently, a number of systems that seek to mimic touch sensitivity have been developed. Although these examples may not fall in the conventional category of polymer mechanochemistry, we felt it was necessary to bring them to light as they can provide an important inspiration for the field and its potential direction. While most of them do not have the ion channels in their transduction process, they seek to follow the same principles of using mechanical perturbations on their surfaces to detect touch. In analyzing the biomimetic systems that have been developed in artificial tactile sensing, Lucarotti *et al.* mainly divided these methods into two approaches, namely synthetic skin and transduction mechanisms, and bio-artificial mechanisms.[51] Bio-artificial approaches are also divided into bio-hybrid systems (synthetic sensors within tissue-engineered skin) and fully biological sensors. These bio-artificial systems often find applications in dermal substitution and repair. Fully biological tactile sensors often consist of an outer layer made from tissue-engineered hydrogel (consisting of cell cultures such as ECM materials) and gelatin containing artificial skin.[51] Using the inspiration from skin cells, Cheneler *et al.* designed a bio-hybrid tactile sensor that is often used to monitor responses of mammalian cell types to normal and tangential loads.[52] In their design (Figure 1.18), they etched microchannels into a silicon substrate to act as a supply route for nutrients to cultured cells that were placed in a well above the substrate and separated by a nanoporous polycarbonate membrane. Conductivity sensors were attached on either side of the well with electrodes and a constant current was kept flowing through the electrodes. Fluctuations in conductance due to extracellular ions caused by applied loads were then detected and analyzed to study the response depending on the load applied.

Synthetic tactile transduction methods often employ capacitative, optoelectric, inductive, piezoelectric and piezoresistive techniques in their design

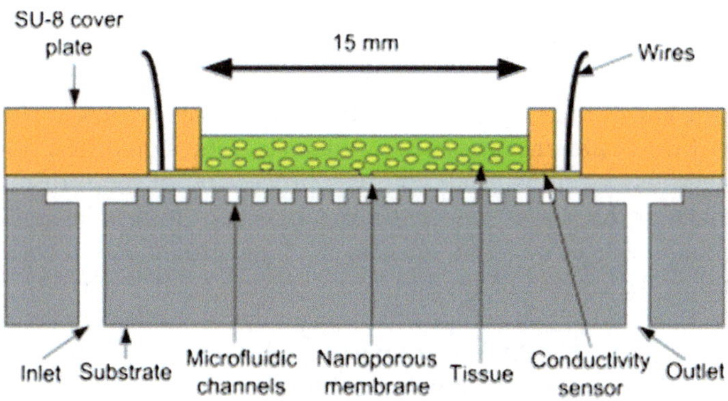

Figure 1.18 Illustration of Cheneler's design of a bio-hybrid tactile sensor. Reprinted from *Microelectronic Engineering*, 97, D. Cheneler *et al.*, Bio-hybrid tactile sensor for the study of the role of mechanoreceptors in human tactile perception, 297–300, Copyright 2012 with permission from Elsevier.

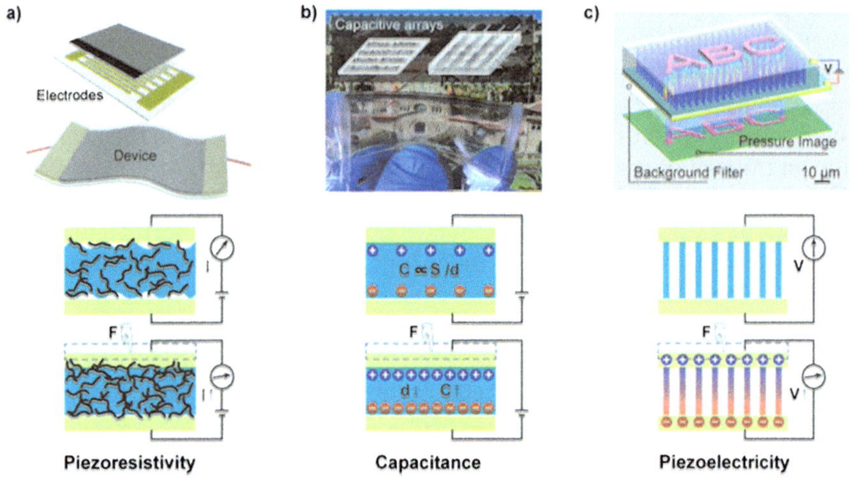

Figure 1.19 Three common transduction methods and their representative devices. (a) Piezoresistivity, (b) capacitance and (c) piezoelectricity. Reproduced with permission from X. Wang *et al.*, *Recent Progress in Electronic Skin*, Advanced Science, John Wiley and Sons, © 2015 The Authors. Published by WILEY-VCH Verlag GmbH & Co. KGaA, Weinheim.

(Figure 1.19). Piezoresistive systems transduce force variations into changes in resistance, which can be detected by simple measuring systems.[53] Piezoelectric systems take advantage of the electrical charges produced when some materials are under a mechanical force in the presence of an electrical dipole moment. Capacitance-based systems monitor changes in the area (A) and distance between electrodes (d) induced by the forces applied to the system. The changes in these dimensions are then used to determine the capacitance ($C = \varepsilon A/d$) based on the material's dielectric constant (ε).

Recently, Jung *et al.* designed a piezoresistive tactile sensor capable of discriminately measuring multidirectional forces.[54] The inspiration for the design of the sensor was based on the organization of mechanoreceptors in the skin. The dermal layer has multiple mechanoreceptors, which allows it to respond to different types of stimulations. The signals are sent to the central nervous system without complex processing of the transduced signals because of the use of different nerve fibers. In Jung's design, the sensor system consists of multiple sensing elements embedded to measure the magnitude and direction of applied forces (Figure 1.20). Each of the sensing elements consisted of two composite films made from PDMS mixed with carbon nanotubes (CNT/PDMS). A conductive adhesive is used to connect the CNT/PDMS films to conductive polymer blocks, which act as electrodes. The force exerted on the contact area reduces the area between two CNT/PDMS films, which in turn causes a change in the resistance of the interlocking films. The sensitivity of the sensing elements was increased by fabricating microdome structures on the film sides facing each other. The changes in the resistance of each element were then used as a measure of the direction

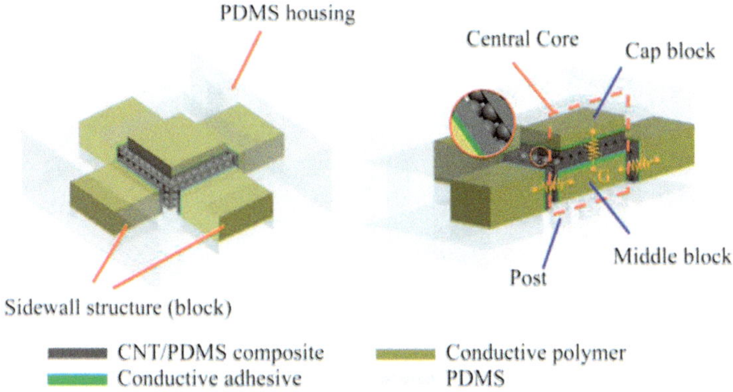

Figure 1.20 Schematic and working principle of a piezoresistive tactile sensor capable of discriminating multidirectional forces. Reproduced from Y. Jung *et al.*, Piezoresistive tactile sensor discriminating multidirectional forces, *Sensors*, **15**(10), 25463–25473.[54]

and magnitudes of the forces applied. The sensor was shown to successfully measure the shear and normal pressures as low as 5.28 kPa and 128 Pa, respectively.[54] In addition to this high sensitivity, the minimal interference was observed between the sensing elements. Results from this open up the possibility of such accurate tactile responsiveness in applications such as slip sensitivity in prosthetic limbs.

The skin's sensory ability has also been increasingly mimicked in the development of the so-called electronic skin (e-skin), which seeks to develop systems which even exceed the resolution and rapid response of the skin itself.[53] The Javey group has been one of the groups that have been producing interesting work in the development of pressure sensitive e-skins.[55,56] They have even developed a user-interactive e-skin that provides instant visual feedback through mapping the applied pressure onto a built-in organic light emitting diode (OLED) matrix.[57] Using transistor arrays, they are able to transduce and amplify the pressure signal into electronic signal. In their design (Figure 1.21), they laminated an OLED matrix between a transistor matrix and a pressure sensitive rubber (PSR). Increases in the pressure applied on the rubber lead to a decrease in resistance by pushing closer embedded nanoparticles that are in the PSR. The decrease in resistance will increase the current flowing through the OLED, thereby increasing the intensity of the output light. This "system-on-plastic" design demonstrates how easy it has become to get instantaneous feedback on artificial tactile systems, something which was not possible only a few decades ago. A significant amount of design in the old systems went into the process of analyzing the data from the mechanoresponse in order to deduce the type of stimuli.

A significant number of these synthetic skin and transduction systems such as those mentioned above have found a variety of applications, from prosthetics, computer touch screens and pressure mapping. In a self-completing cycle, some of these wearable artificial tactile sensors are now being

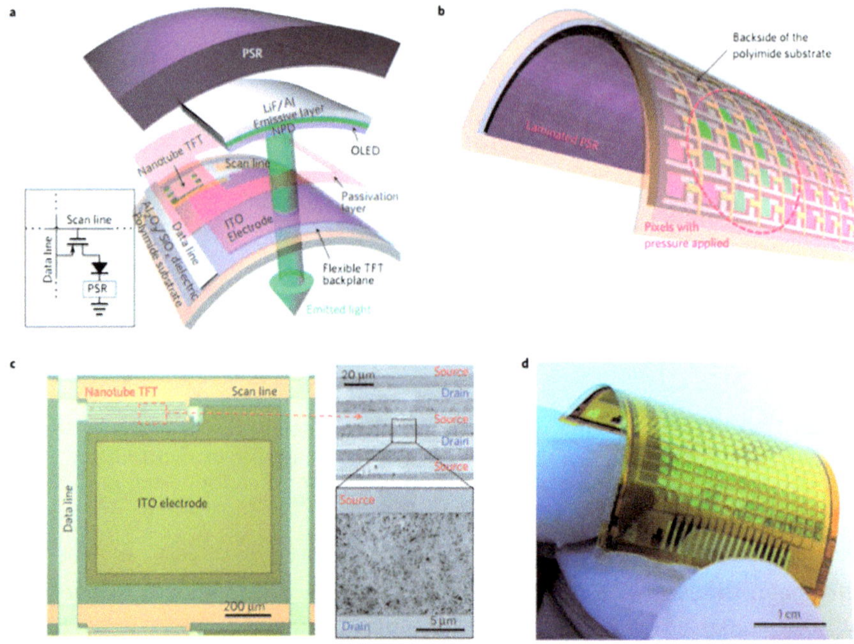

Figure 1.21 Concept and structure of the user-interactive e-skin. a. Schematic layout of a single pixel, consisting of a nanotube TFT, an OLED and a pressure sensor (PSR) integrated vertically on a polyimide substrate. b. Schematic diagram of an array of pixels (16 × 16) functioning as an interactive e-skin, capable of spatially mapping and visually displaying an applied pressure profile. c. Optical micrograph of a fabricated pixel before the integration of the OLED and PSR. The drain of the TFT is connected to an ITO pad, which serves as the anode electrode for the corresponding OLED. Scanning electron micrographs of the active area of a nanotube TFT are also shown. d. Optical photograph of a fully fabricated interactive e-skin containing 16 × 16 pixels with a size of ~3 × 3.5 cm^2. Reprinted by permission from Macmillan Publishers Ltd: *Nature Materials* (ref. 53). Copyright 2013.

used to monitor the skin – using its own technique to monitor it.[58] While power consumption has been a traditional deterrent in the use of some of these systems, the emergence of better design techniques has led to more energy-efficient systems. In fact, some systems are now able to self-power, with flexible solar cells[59] and piezoelectric nanogenerators[60] recently gaining prominence. Wang *et al.* have a good review that discusses some of the most recent and exciting developments in the development of e-skin.[53]

1.4 Conclusion

In this chapter, the role of mechanical forces in a variety of biological systems was highlighted. The conversion of these mechanical forces into chemical bonds is also important. A number of mechanisms and mechanotransduction pathways in a variety of systems have also been highlighted, providing inspiration on how the complex biological systems can provide ideas to design systems that serve as remedies for when these systems are malfunctioning or for other systems altogether. Some of the ways in which these mechano-based systems have been mimicked by other scientists have also been highlighted.

The importance of mechanical forces was shown in the bone development, from growth to repair, and the effect of mechanical forces in space exploration was also highlighted. We also got a glimpse of some applications that are mainly based on the bones such as self-healing systems. Work by Discher *et al.* showed the importance of mechanical forces in exposing binding sites in folded proteins. This work has given the idea of cryptic sites, which were exploited by Jierry and coworkers in the developing field of soft mechanochemistry, where the use of small mechanical forces is exposing these cryptic sites and leading to chemical changes with larger energy requirements. The effects of failures in mechanotransduction pathways in living organisms was also observed, with a long list of diseases associated with defects in mechanotransduction, from cancer, central nervous system disorders and immune system disorders. Such cases provide challenges to polymer chemists to use this knowledge of various mechanotransduction pathways in order to develop materials that can perform the same functions to act as remedies or outright replacements. Lastly, the effect of mechanical forces on the sensitivity in both hair and skin was also discussed. The mechanoresponsiveness and sensitivity have now been mimicked to develop systems with cochlear and tactile sensing abilities that can serve not only as replacements in the case of permanent injuries and genetic defects, but in other applications such as pressure mapping, energy production and even humanoid skin. Such work can serve as a reminder that Nature is always an endless source of inspiration.

References

1. (a) W. Ostwald, Die chemische Literatur und die Organisation der Wissenschaft, in *Handbuch der allgemeinen Chemie*, ed. W. Ostwald and C. Drucker, Akademische Verlagsgesellschaft m. b. H., Leipzig, 1919, p. 70; (b) Die chemische Literatur und die Organisation der Wissenschaft, in *Handbuch der allgemeinen Chemie*, ed. W. Ostwald and C. Drucker, Akademische Verlagsgesellschaft m. b. H., Leipzig, 1919, p. 77.
2. K. D. M. Harris, Mechanochemical synthesis: How grinding evolves, *Nat. Chem.*, 2013, **5**(1), 12–14.
3. R. A. Haley, J. Mack and H. Guan, 2-in-1: catalyst and reaction medium, *Inorg. Chem. Front.*, 2017, **4**(1), 52–55.
4. M. Reibold, P. Paufler, A. A. Levin, W. Kochmann, N. Patzke and D. C. Meyer, Materials: carbon nanotubes in an ancient Damascus sabre, *Nature*, 2006, **444**(7117), 286.
5. *Where Are Composites Used?*, American Composites Manufacturers Association, available from: http://www.acmanet.org/composites/where-are-composites-used.

6. H. W. Melville and A. J. R. Murray, The ultrasonic degradation of polymers, *Trans. Faraday Soc.*, 1950, **46**, 996–1009.
7. H. Staudinger and E. O. Leupold, Über Isopren und Kautschuk, 18. Mitteil.: Viscositäts-Untersuchungen an Balata, *Ber. Dtsch. Bot. Ges. (Ser. A Ser. B)*, 1930, **63**(3), 730–733.
8. J. Li, C. Nagamani and J. S. Moore, Polymer Mechanochemistry: From Destructive to Productive, *Acc. Chem. Res.*, 2015, **48**(8), 2181–2190.
9. N. R. Council, *Hierarchical Structures in Biology as a Guide for New Materials Technology*, National Academies Press, 1994.
10. T. M. Hermans, H. Frauenrath and F. Stellacci, Droplets out of Equilibrium, *Science*, 2013, **341**(6143), 243–244.
11. D. E. Jaalouk and J. Lammerding, Mechanotransduction gone awry, *Nat. Rev. Mol. Cell Biol.*, 2009, **10**(1), 63–73.
12. S. Majkut, P. C. Dingal and D. E. Discher, Stress sensitivity and mechanotransduction during heart development, *Curr. Biol.*, 2014, **24**(10), R495–R501.
13. A. L. Black Ramirez, Z. S. Kean, J. A. Orlicki, M. Champhekar, S. M. Elsakr and W. E. Krause, *et al.*, Mechanochemical strengthening of a synthetic polymer in response to typically destructive shear forces, *Nat. Chem.*, 2013, **5**(9), 757–761.
14. A. W. Orr, B. P. Helmke, B. R. Blackman and M. A. Schwartz, Mechanisms of mechanotransduction, *Dev. Cell*, 2006, **10**(1), 11–20.
15. P. Fratzl and R. Weinkamer, Nature's hierarchical materials, *Prog. Mater. Sci.*, 2007, **52**(8), 1263–1334.
16. J. Chalmers and R. D. Ray, The Growth of Transplanted Foetal Bones in Different Immunological Environments, *J. Bone Jt. Surg., Br. Vol.*, 1962, **44-B**(1), 149–164.
17. N. A. Sims and T. J. Martin, Coupling the activities of bone formation and resorption: a multitude of signals within the basic multicellular unit, *BoneKEy Rep.*, 2014, **3**, 481.
18. Y. Mawani, J. F. Cawthray, S. Chang, K. Sachs-Barrable, D. M. Weekes and K. M. Wasan, *et al., In vitro* studies of lanthanide complexes for the treatment of osteoporosis, *Dalton Trans.*, 2013, **42**(17), 5999–6011.
19. H. M. Frost, Bone "mass" and the "mechanostat": A proposal, *Anat. Rec.*, 1987, **219**(1), 1–9.
20. S. R. White, N. R. Sottos, P. H. Geubelle, J. S. Moore, M. R. Kessler and S. R. Sriram, *et al.,* Autonomic healing of polymer composites, *Nature*, 2001, **409**(6822), 794–797.
21. G. R. Gossweiler, G. B. Hewage, G. Soriano, Q. Wang, G. W. Welshofer and X. Zhao, *et al.,* Mechanochemical Activation of Covalent Bonds in Polymers with Full and Repeatable Macroscopic Shape Recovery, *ACS Macro Lett.*, 2014, **3**(3), 216–219.
22. K. Imato, T. Kanehara, S. Nojima, T. Ohishi, Y. Higaki and A. Takahara, *et al.,* Repeatable mechanochemical activation of dynamic covalent bonds in thermoplastic elastomers, *Chem. Commun.*, 2016, **52**(69), 10482–10485.

23. B. A. Beiermann, S. L. B. Kramer, P. A. May, J. S. Moore, S. R. White and N. R. Sottos, The Effect of Polymer Chain Alignment and Relaxation on Force-Induced Chemical Reactions in an Elastomer, *Adv. Funct. Mater.*, 2014, **24**(11), 1529–1537.
24. J. R. Mosley and L. E. Lanyon, Strain rate as a controlling influence on adaptive modeling in response to dynamic loading of the ulna in growing male rats, *Bone*, 1998, **23**(4), 313–318.
25. R. Ruimerman, P. Hilbers, B. van Rietbergen and R. Huiskes, A theoretical framework for strain-related trabecular bone maintenance and adaptation, *J. Biomech.*, 2005, **38**(4), 931–941.
26. M. N. Holme, I. A. Fedotenko, D. Abegg, J. Althaus, L. Babel and F. Favarger, *et al.*, Shear-stress sensitive lenticular vesicles for targeted drug delivery, *Nat. Nanotechnol.*, 2012, **7**(8), 536–543.
27. T. Saxer, A. Zumbuehl and B. Müller, The use of shear stress for targeted drug delivery, *Cardiovasc. Res.*, 2013, **99**(2), 328–333.
28. S. R. White, J. S. Moore, N. R. Sottdos, B. P. Krull, W. A. S. Cruz and R. C. R. Gergely, Restoration of Large Damage Volumes in Polymers, *Science*, 2014, **344**(6184), 620–623.
29. B. J. Blaiszik, S. L. B. Kramer, S. C. Olugebefola, J. S. Moore, N. R. Sottos and S. R. White, Self-Healing Polymers and Composites, *Annu. Rev. Mater. Res.*, 2010, **40**(1), 179–211.
30. K. S. Toohey, N. R. Sottos, J. A. Lewis, J. S. Moore and S. R. White, Self-healing materials with microvascular networks, *Nat. Mater.*, 2007, **6**(8), 581–585.
31. Y. Yang and M. W. Urban, Self-healing polymeric materials, *Chem. Soc. Rev.*, 2013, **42**(17), 7446–7467.
32. A. Lavrenova, J. Farkas, C. Weder and Y. C. Simon, Visualization of Polymer Deformation Using Microcapsules Filled with Charge-Transfer Complex Precursors, *ACS Appl. Mater. Interfaces*, 2015, **7**(39), 21828–21834.
33. M. J. Robb, W. Li, R. C. R. Gergely, C. C. Matthews, S. R. White and N. R. Sottos, *et al.*, A Robust Damage-Reporting Strategy for Polymeric Materials Enabled by Aggregation-Induced Emission, *ACS Cent. Sci.*, 2016, **2**(9), 598–603.
34. B. Di Credico, G. Griffini, M. Levi and S. Turri, Microencapsulation of a UV-Responsive Photochromic Dye by Means of Novel UV-Screening Polyurea-Based Shells for Smart Coating Applications, *ACS Appl. Mater. Interfaces*, 2013, **5**(14), 6628–6634.
35. C. E. Diesendruck, N. R. Sottos, J. S. Moore and S. R. White, Biomimetic Self-Healing, *Angew. Chem., Int. Ed.*, 2015, **54**(36), 10428–10447.
36. P. Lavalle, F. Boulmedais, P. Schaaf and L. Jierry, Soft-Mechanochemistry: Mechanochemistry Inspired by Nature, *Langmuir*, 2016, **32**(29), 7265–7276.
37. M. V. Tirrell, *Reaction Kinetics and Structure of Enzyme Macromolecules Subjected to Hydrodynamic Shearing Forces: Enzyme Mechanochemistry*, 1977.

38. J. H. Collier, J. S. Rudra, J. Z. Gasiorowski and J. P. Jung, Multi-component extracellular matrices based on peptide self-assembly, *Chem. Soc. Rev.*, 2010, **39**(9), 3413–3424.
39. C. P. Johnson, H. Y. Tang, C. Carag, D. W. Speicher and D. E. Discher, Forced unfolding of proteins within cells, *Science*, 2007, **317**(5838), 663–666.
40. K. A. Beck and W. J. Nelson, A spectrin membrane skeleton of the Golgi complex, *Biochim. Biophys. Acta, Mol. Cell Res.*, 1998, **1404**(1–2), 153–160.
41. C. Rios, J. Longo, S. Zahouani, T. Garnier, C. Vogt and A. Reisch, *et al.*, A new biomimetic route to engineer enzymatically active mechano-responsive materials, *Chem. Commun.*, 2015, **51**(26), 5622–5625.
42. L. Loufrani, C. Dubroca, D. You, Z. Li, B. Levy and D. Paulin, *et al.*, Absence of Dystrophin in Mice Reduces NO-Dependent Vascular Function and Vascular Density: Total Recovery After a Treatment with the Aminoglycoside Gentamicin, *Arterioscler., Thromb., Vasc. Biol.*, 2004, **24**(4), 671–676.
43. P. A. Fuchs, How many proteins does it take to gate hair cell mechanotransduction? *Proc. Natl. Acad. Sci. U. S. A.*, 2015, **112**(5), 1254–1255.
44. B. Zhao and U. Muller, The elusive mechanotransduction machinery of hair cells, *Curr. Opin. Neurobiol.*, 2015, **34**, 172–179.
45. A. Hudspeth and D. Corey, Sensitivity, polarity, and conductance change in the response of vertebrate hair cells to controlled mechanical stimuli, *Proc. Natl. Acad. Sci.*, 1977, **74**(6), 2407–2411.
46. T. Hoffmann, K. M. Tych, M. L. Hughes, D. J. Brockwell and L. Dougan, Towards design principles for determining the mechanical stability of proteins, *Phys. Chem. Chem. Phys.*, 2013, **15**(38), 15767–15780.
47. M. Asadnia, A. G. Kottapalli, K. D. Karavitaki, M. E. Warkiani, J. Miao and D. P. Corey, *et al.*, From Biological Cilia to Artificial Flow Sensors: Biomimetic Soft Polymer Nanosensors with High Sensing Performance, *Sci. Rep.*, 2016, **6**, 32955.
48. C. M. Walsh, D. M. Bautista and E. A. Lumpkin, Mammalian touch catches up, *Curr. Opin. Neurobiol.*, 2015, **34**, 133–139.
49. E. A. Lumpkin and M. J. Caterina, Mechanisms of sensory transduction in the skin, *Nature*, 2007, **445**(7130), 858–865.
50. E. A. Lumpkin, K. L. Marshall and A. M. Nelson, The cell biology of touch, *J. Cell Biol.*, 2010, **191**(2), 237–248.
51. C. Lucarotti, C. M. Oddo, N. Vitiello and M. C. Carrozza, Synthetic and bio-artificial tactile sensing: a review, *Sensors (Basel)*, 2013, **13**(2), 1435–1466.
52. D. Cheneler, M. C. L. Ward and C. J. Anthony, Bio-hybrid tactile sensor for the study of the role of mechanoreceptors in human tactile perception, *Microelectron. Eng.*, 2012, **97**, 297–300.
53. X. Wang, L. Dong, H. Zhang, R. Yu, C. Pan and Z. L. Wang, Recent Progress in Electronic Skin, *Adv. Sci. (Weinh.)*, 2015, **2**(10), 1500169.
54. Y. Jung, D. G. Lee, J. Park, H. Ko and H. Lim, Piezoresistive Tactile Sensor Discriminating Multidirectional Forces, *Sensors (Basel)*, 2015, **15**(10), 25463–25473.

55. K. Takei, T. Takahashi, J. C. Ho, H. Ko, A. G. Gillies and P. W. Leu, *et al.*, Nanowire active-matrix circuitry for low-voltage macroscale artificial skin, *Nat. Mater.*, 2010, **9**(10), 821–826.
56. T. Takahashi, K. Takei, A. G. Gillies, R. S. Fearing and A. Javey, Carbon nanotube active-matrix backplanes for conformal electronics and sensors, *Nano Lett.*, 2011, **11**(12), 5408–5413.
57. C. Wang, D. Hwang, Z. Yu, K. Takei, J. Park and T. Chen, *et al.*, User-interactive electronic skin for instantaneous pressure visualization, *Nat. Mater.*, 2013, **12**(10), 899–904.
58. S. Lim, D. Son, J. Kim, Y. B. Lee, J.-K. Song and S. Choi, *et al.*, Transparent and Stretchable Interactive Human Machine Interface Based on Patterned Graphene Heterostructures, *Adv. Funct. Mater.*, 2015, **25**(3), 375–383.
59. M. Kaltenbrunner, M. S. White, E. D. Głowacki, T. Sekitani, T. Someya and N. S. Sariciftci, *et al.*, Ultrathin and lightweight organic solar cells with high flexibility, *Nat. Commun.*, 2012, **3**, 770.
60. K. C. Pradel, W. Wu, Y. Ding and Z. L. Wang, Solution-Derived ZnO Homojunction Nanowire Films on Wearable Substrates for Energy Conversion and Self-Powered Gesture Recognition, *Nano Lett.*, 2014, **14**(12), 6897–6905.

CHAPTER 2

Mechanophores for Chemical Function

C. E. DIESENDRUCK

Technion – Israel Institute of Technology, Schulich Faculty of Chemistry, Technion City, Haifa, 320008, Israel
*E-mail: charles@tx.technion.ac.il

2.1 Introduction

As described in other chapters, mechanical energy can be used to overcome a transition barrier and drive a chemical reaction.[1] However, the force applied not only provides energy to the molecule to bypass a barrier, it also affects the barrier, the energy of reactants and products, as well as all the thermodynamic parameters of competing reactions, including some that are typically forbidden or provide low yields. In this chapter, we focus on how the mechanochemistry principles described in other chapters were applied to use mechanical energy in order to induce different chemical reactions. These reactions can be generally divided into subgroups according to the goal of the mechanophore: creation of reactive species, latent catalyst activation, spectral change, production of small molecules and others.[2] Each of these groups are described separately, including some mechanophores that present more than one chemical change.

Polymer Chemistry Series No. 26
Mechanochemistry in Materials
Edited by Yoan C. Simon and Stephen L. Craig

Published by the Royal Society of Chemistry, www.rsc.org

2.2 Creation of Reactive Species

2.2.1 Radicals

Bond scission is the most common mechanochemical reaction. When stress is applied to polymeric materials, in addition to chain slipping, transformations occur in the molecular level, such as changes in dihedral angles, bond angles and bond distances.[3] Due to alterations in the bell curve of a chemical bond,[4] bond scission may occur, typically leading to two macroradicals (homolytic bond scission).

Covalent bond scission in polymeric materials has been known since the 1930s. After a long struggle to have the idea of covalent macromolecules accepted by the academic community[5] (as opposed to polymers being aggregates of smaller molecules), Staudinger ball-milled high molecular weight polystyrene and observed the reduction of its average molecular weight.[6] Posterior experimental studies pointed to homolytic and/or heterolytic mechanisms in typical homopolymers, but the leading argument was provided in solution studies carried out by Sohma *et al.* using electron spin resonance (ESR).[7] Their results showed that even without mechanophores, most homopolymers undergo homolytic bond scission, producing reactive radicals that can be observed experimentally by ESR or using spin traps.

The first synthetic "mechanophore" produced radicals as well. Encina's seminal work on peroxide containing polyvinylpyrrolidones demonstrated that mechanochemical bond scission events were accelerated and selective if weaker bonds were incorporated into the main chain of a polymer.[8] Peroxide bonds (–O–O–) broke faster than carbon–carbon bonds, producing terminal oxygen radicals.

Additional mechanophores were developed based on the simple principles developed by Encina. Moore *et al.* demonstrated one of the first applications of this approach by preparing polymers with chain-centered azo isobutyronitriles groups.[9] Upon stress, a C–N bond undergoes homolytic bond scission, which is followed by release of N_2 producing two carbon radicals, which are stabilized by the cyano groups. To further understand the products of the reaction, the mechanophore was labeled with ^{13}C and the sonication products analyzed by ^{13}C-NMR. Interestingly, after sonication, cyanohydrin was obtained as product, in contrast to the products obtained by thermolysis of the azo bond (Scheme 2.1).

Otsuka *et al.* developed a mechanophore that produces even more stable carbon radicals based on diarylbibenzofuranone (DABBF).[10] The sigma bond of DABBF is weak and even at room temperature is in equilibrium with the radical species. Applying force to a polymer that contains this mechanophore pushes the equilibrium towards the radical product (Figure 2.1). The radical product not only is relatively stable to oxygen, but its blue color can also be used to measure bond scission (in addition to EPR). Indeed, this mechanophore was used by the Otsuka group to demonstrate again that stress in the freezing of polymer gels leads to bond scission.

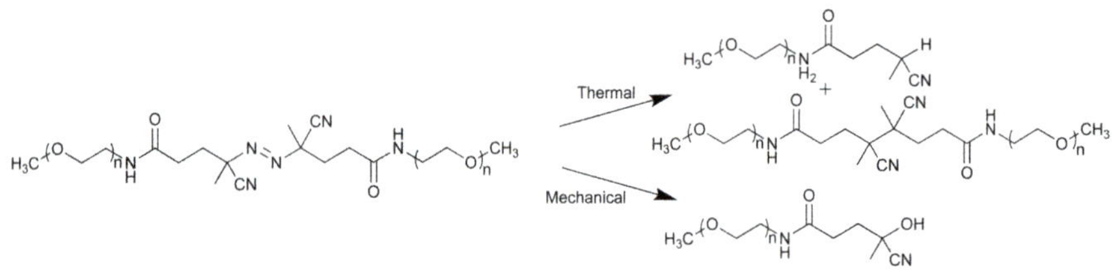

Scheme 2.1 Products of thermal and mechanochemical activation of azo isobutyronitriles.

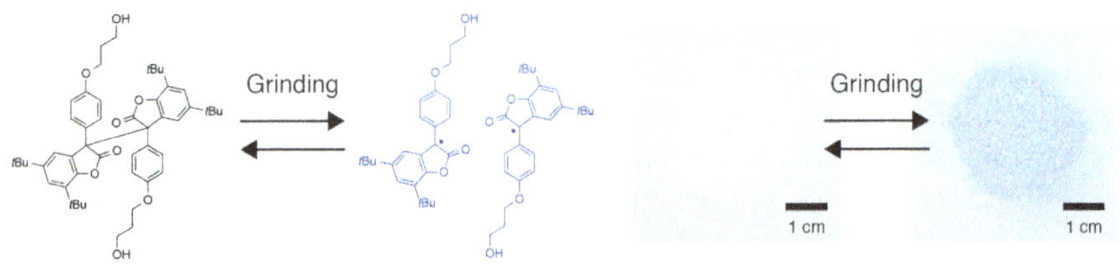

Figure 2.1 Mechanochemical homolytic scission of DABBF. Reproduced from ref. 10 with permission from John Wiley and Sons. Copyright © 2015 WILEY-VCH Verlag GmbH & Co. KGaA, Weinheim.

By using disulfide bonds, Sheiko *et al.* developed an additional mechanophore that undergoes selective homolytic bond scission.[11] Polysulfide bonds in vulcanized rubber were previously known to undergo selective bond scission and provided self-healing capabilities to rubber in the absence of oxygen.[12] Sheiko used this mechanophore to study the self-generated tension in densely grafted bottlebrush polymers. Interestingly, these forces are so strong that they lead not only to S–S bond scission, but also to C–C bond scission. More recently, Sijbesma *et al.*, while trying to induce the mechanochemical retro [2 + 2] of a four-membered ketene dimer ring, noticed that benzylic C–O bonds undergo faster bond scission, producing terminal carbon and oxygen radicals instead of the desired ketenes.[13]

Craig *et al.* have tested difluorocyclopropanes as mechanophores.[14] In contrast with the dichloro- and dibromocyclopropanes (described below), these do not undergo electrocyclic ring-opening, but undergo homolytic bond scission to produce carbon radicals at close proximity, which rebind and reform the cyclopropane after the strain is released. However, if stress is maintained, the radicals are stable and can be trapped by other reagents. Interestingly, if two of these trapped radicals are adjacent, an additional homolytic C–C bond scission occurs between them producing polymers with terminal double bonds (Table 2.1).[15]

2.2.2 Carbon Cations

Heterolytic bond scission in mechanochemistry of polymeric materials is a much rarer event than homolytic bond scission. Sakaguchi observed heterolytic bond scission in several homopolymers, both polar[16] and non-polar,[17]

Table 2.1 Mechanophores that produce radicals.

Mechanophore	Radicals produced	Ref.
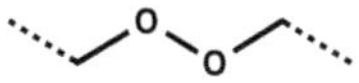	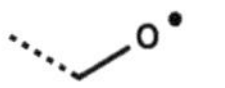	8
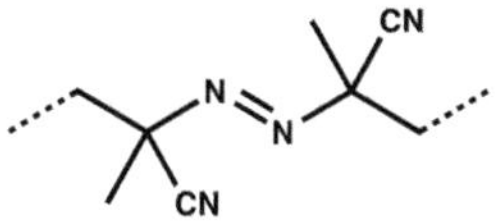	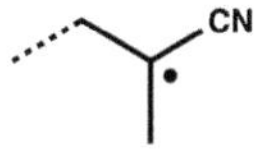	9
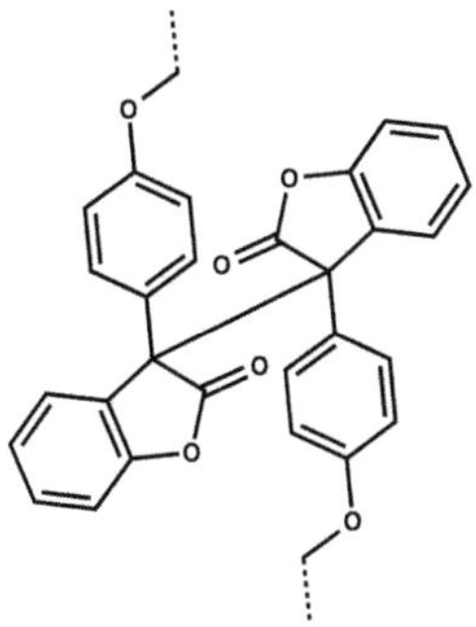	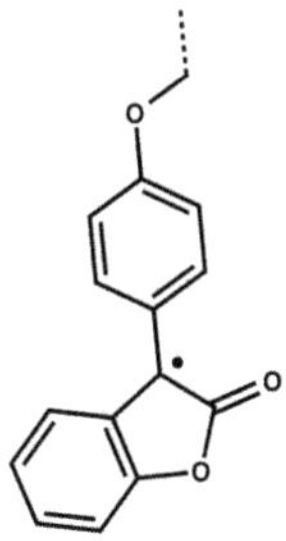	10
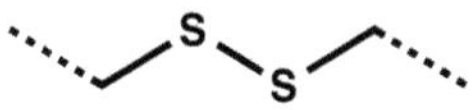		11
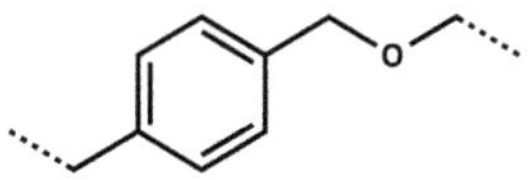	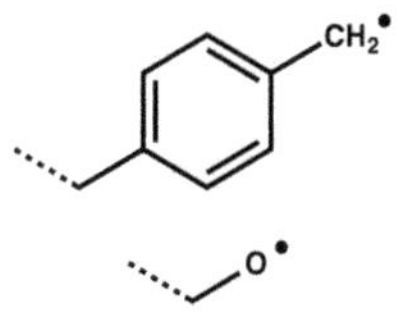	13
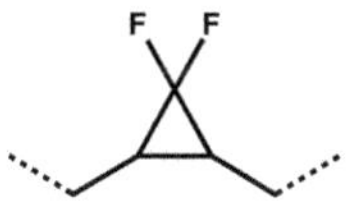	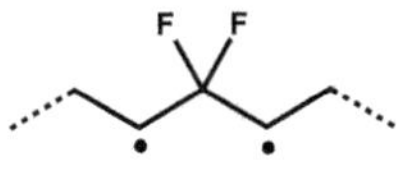	14

using anion traps by ESR. However, Sohma's results[7] as well as theoretical studies disagreed with Sakaguchi's observations. Still, in some polymers made by very polar bonds, such as PDMS or silicate glass, heterolytic bond scission is still expected, producing cations and anions.[18]

Along these lines, a triaryl sulfonium salt was tested as a mechanophore, expecting bond scission at the long C–S bond.[19] Theoretical calculations indicated that after bond scission, instead of the expected homolytic bond scission, a thioether and phenyl cation are observed, resulting from heterolytic bond scission. The phenyl cation could be trapped with an amine, but, if no trap was present, the phenyl cation reacted with the triflate counteranion in an unexpected reaction, which was later demonstrated in solution (Table 2.2).[20]

2.2.3 Reactive Organic Functional Groups

Several mechanophores were prepared with the goal of inducing bond-forming reactions after mechanochemical scission. A successful mechanophore could lead to effective self-reinforcing materials or even self-healing

Table 2.2 Mechanophores that produce reactive cations.

Mechanophore	Cation produced	Ref.
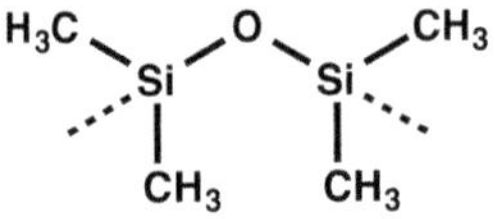	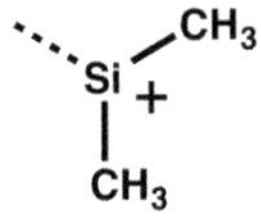	18
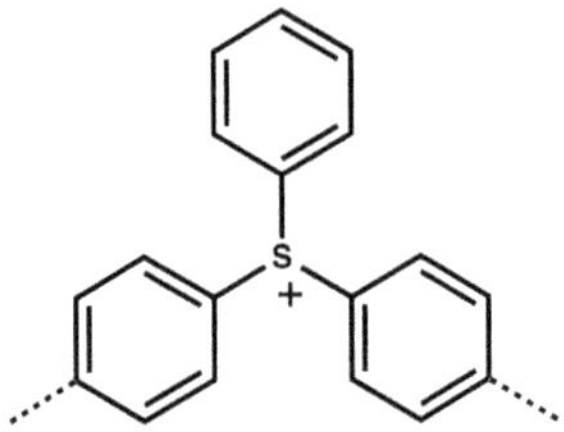	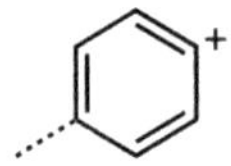	19

materials, where after chemical bonds are broken, new chemical bonds are formed to restore the original properties.[21]

Moore *et al.* developed this idea by carrying out a mechanochemical electrocyclic ring-opening of benzocyclobutenes in order to produce reactive dearomatized *ortho*-quinodimethide.[22] This species reacts rapidly at room temperature with dienophiles, driven by rearomatization. Unexpectedly, the authors found out that the stereochemistry of the reaction doesn't follow traditional Woodward–Hoffman rules. Instead, the ring-opening always affords the *E*,*E* product, following a simple "pulling" logic, *i.e.* starting from either *cis* or *trans* isomers, the product is always the same (Figure 2.2). Many years later, Craig *et al.* used this mechanophore successfully in mechanochemical bond-forming reactions in polymers under ultrasonication.[23] The *ortho*-quinodimethide produced reacted with added bis-dienophiles, leading to increased molecular weight or cross-linking.

Moore *et al.* tested dicyanocyclobutanes as mechanophores, which under mechanical stress underwent selective retro [2 + 2] cycloaddition to produce cyanoacrylates.[24] Cyanoacrylates are extremely reactive Michael acceptors, which can be immediately trapped using nucleophiles such as secondary amines. The number of substituents in the cyclobutane and their relative stereochemistry were observed to have a significant influence in the rate of mechanochemical reaction.[25] Craig *et al.* improved the mechanochemical production of Michael acceptors by using a substituted bicyclo[3.2.0]heptane that contained the parent cyclobutane.[26] Upon mechanochemical activation, the main chain does not necessarily undergo scission and the obtained dienone is not only reactive toward nucleophiles, but it can also be re-cyclized into the original mechanophore using light. Craig *et al.* further improved this mechanophore using bicyclo[4.2.0]octane, increasing the released strain.[27] Recently, Moore *et al.* further increased the portfolio of reactivities accessible by mechanochemistry, producing a ketene through a similar retro-Staudinger cycloaddition.[28] The first demonstration that these four-membered cycles could be used in mending[29] was shown by Craig, using

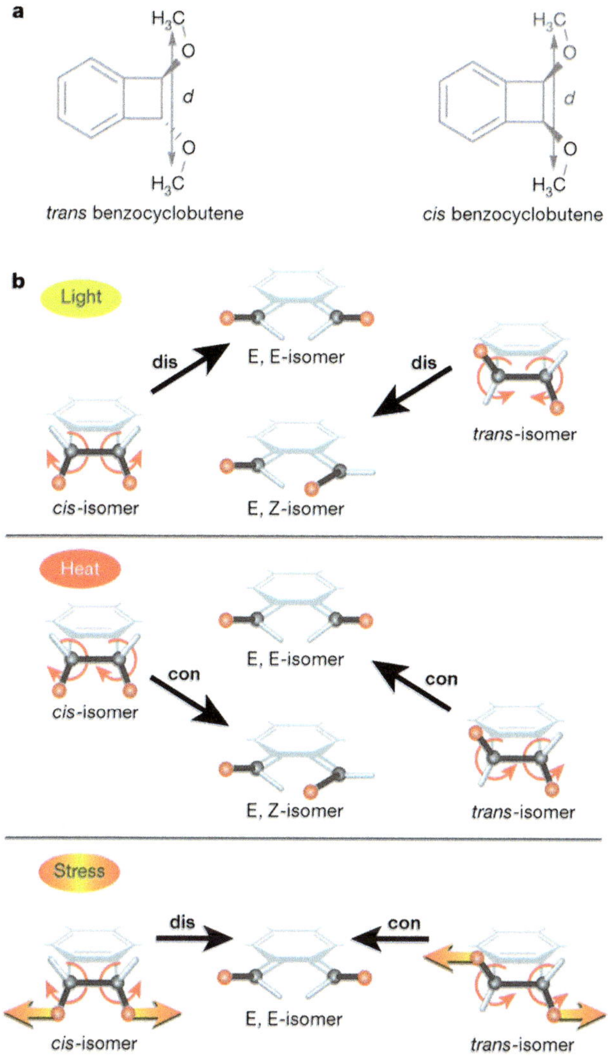

Figure 2.2 (a) Different isomers of benzocyclobutenes and (b) their light, heat and mechanically induced ring-opening. Reprinted by permission from Macmillan Publishers Ltd: *Nature*. (ref. 22) Copyright 2007.

perfluorocyclobutane aryl ethers as mechanophores.[30] Like the cyclic mechanophores described above, perfluorocyclobutane also undergoes retro [2 + 2] cycloaddition under stress, breaking the polymer into smaller segments with stable trifluorovinyl ether terminal groups. Upon heating of these segments, the trifluorovinyl ether groups underwent cycloaddition, and the average molecular weight of the polymer mixture increased again.

Craig *et al.* demonstrated that even changing the reactivity of a functional group could be enough to induce a bond-forming reaction. Dibromocyclopropanes, under mechanical stress, undergo electrocyclic ring opening to

2,3-dibromoalkenes.[31] The allylic bromide is more reactive to nucleophilic substitution than the original dibromocyclopropane. The authors demonstrated that sebacate nucleophiles added to the polymer led to stress-induced cross-linking, effectively producing a self-reinforcing material.[32] Craig *et al.* also tested epoxides as mechanophores.[33] In contrast with the dibromocyclopropanes, these undergo scission of the C–C bond to form an ylide that can be trapped with dipolarophiles or alcoholic solvents, producing larger cyclic ethers or non-symmetric acetals, respectively. These mechanophores are harder to trigger and were only active in the presence of an adjacent cyclic alkane. This provided the first demonstration of levers in mechanochemistry – structural changes close to the mechanophore that increase their ease or rate of activation.[34]

Recently, Moore *et al.* have shown that bond scission is not always required to induce a change in reactivity. Internal alkynes in polymers were shown by theoretical calculations to bend under stress reducing the 180° dihedral angle and so leading to the stabilization of the π^* orbital. Indeed, a poly(methyl acrylate) having an internal alkyne was shown to react with azide nucleophiles even before bond scission (Table 2.3).[35]

2.3 Catalyst Activation

Metal–ligand bonds are typically long non-covalent bonds and therefore good candidates to produce mechanophores. Naturally, if the ligand sphere of a metal is changed, its catalytic activity is affected.[36] Sijbesma *et al.* showed that the Pd–P bond could be mechanically broken and thermally reformed with time.[37] In that article, the authors already indicated the potential for using mechanochemical reactions to open a coordination site for catalysis.

Sijbesma *et al.* then developed a polyTHF-substituted N-heterocyclic carbene (NHC) ligand, which coordinates strongly to metals and acts as a strong σ donator. When two NHC ligands coordinate a metal, the metal–ligand bond is at the center of the linear polymer and, therefore, is susceptible to mechanochemical bond scission.[38]

Sijbesma demonstrated the mechanochemical activation of latent olefin-metathesis catalysts using this ligand system.[39] The precatalyst with two NHC ligands is inactive since the NHC ligand is not labile under ambient conditions. Ultrasonication provides enough energy to break the NHC–Ru bond, opening a vacancy for the alkene to coordinate and the reaction to start through the formation of a metallocyclobutane and subsequent metathesis reactions. The mechanoactivated catalyst is capable of inducing ring-closing metathesis or even ring-opening metathesis polymerization.[40] The NHC ligand itself can also work as an organocatalyst. The carbene is unreactive when coordinating a metal such as silver, but upon mechanochemical activation the NHC is released to solution catalyzing, for example, the transesterification of esters (Table 2.4, Scheme 2.2).[41]

Table 2.3 Mechanophores that produce reactive groups.

Mechanophore	Reactive Group Produced	Ref.
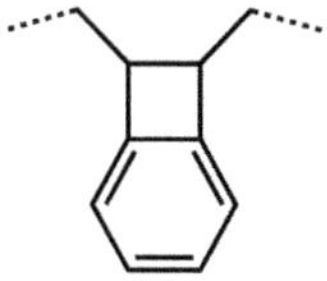	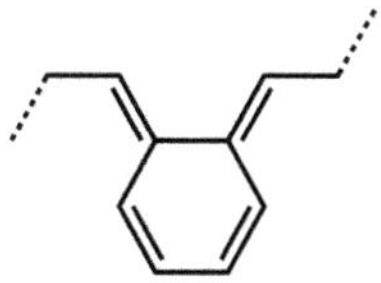	22,23
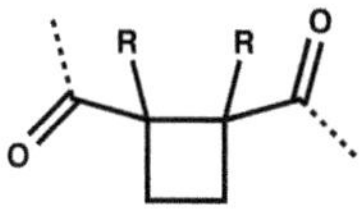 R = H, CN	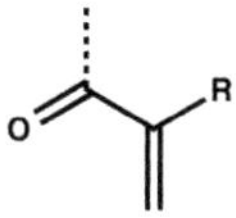 R = H, CN	24,25
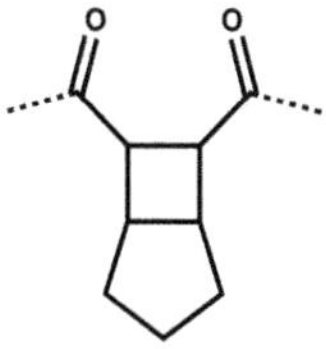	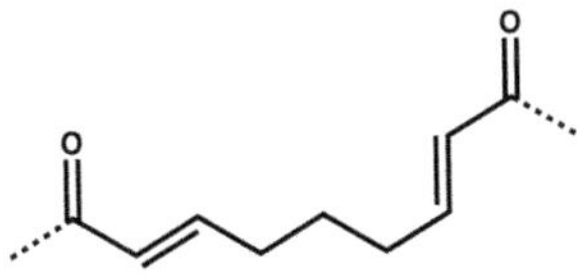	26
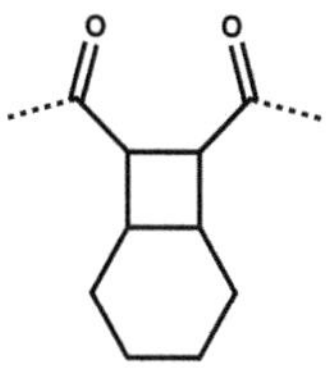	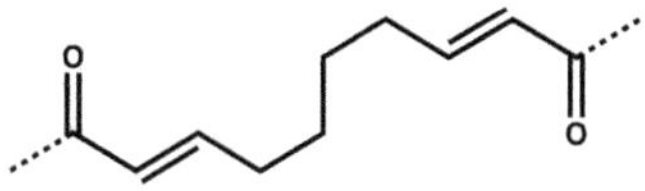	27
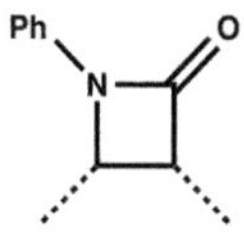		28
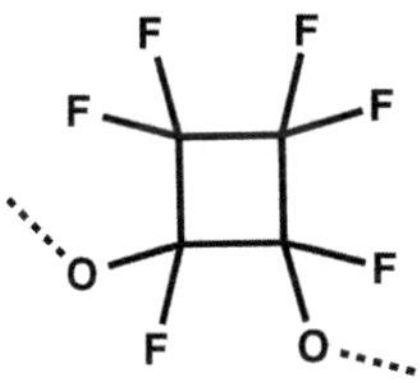	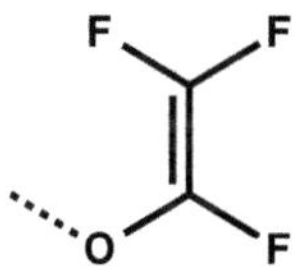	30
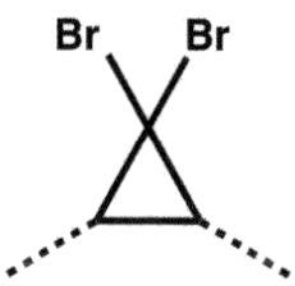	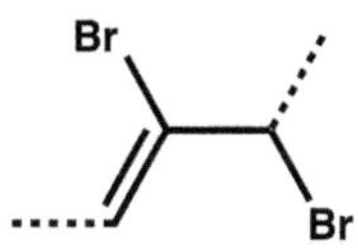	32
	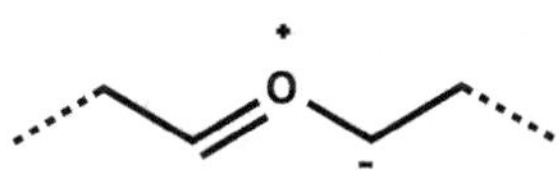	33
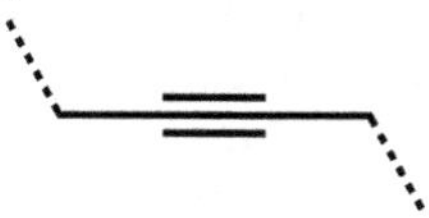	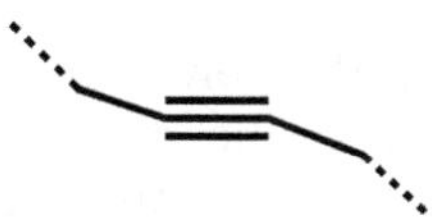	35

Table 2.4 Latent catalysts as mechanophores.

Mechanophore	Catalysis activated	Ref.
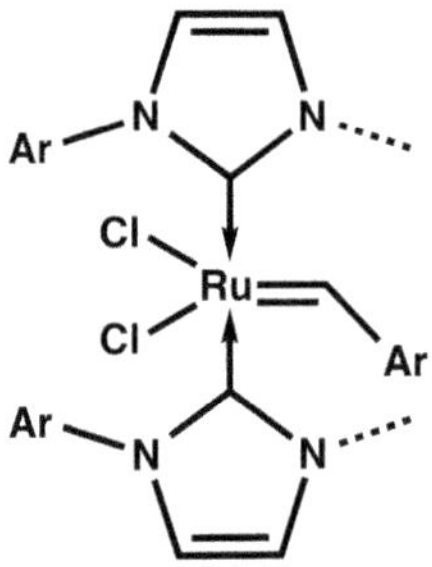	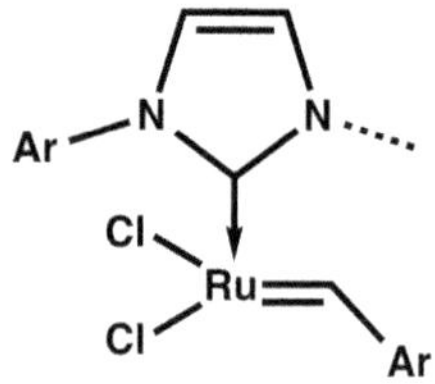	39,40
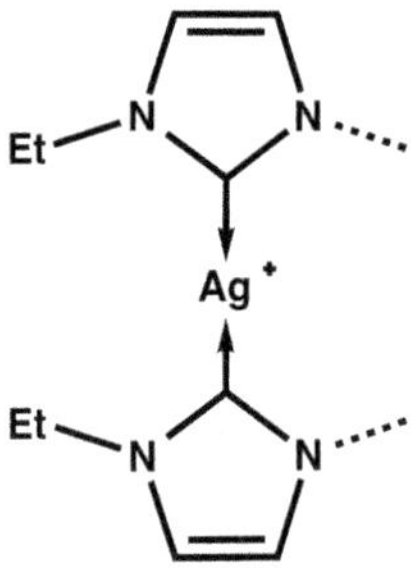	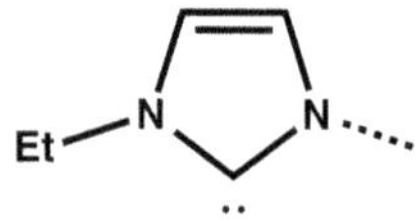	41

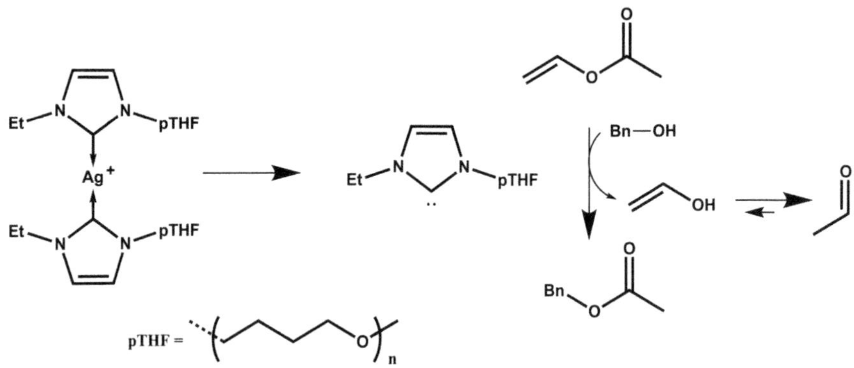

Scheme 2.2 Mechanochemically triggered organocatalytic transesterification.

2.4 Spectral Change

For absorption and luminescence spectral changes to occur, a bond scission event is not always required. Simply bending bonds can be enough to disrupt long conjugations. This was shown three decades ago in partially cross-linked polydiacetylenes.[42] When the polymer phase was ordered, conjugation was maximized. However, when the material was stretched, the absorbance diminished in intensity and blue-shifted due to the breaking of conjugation.

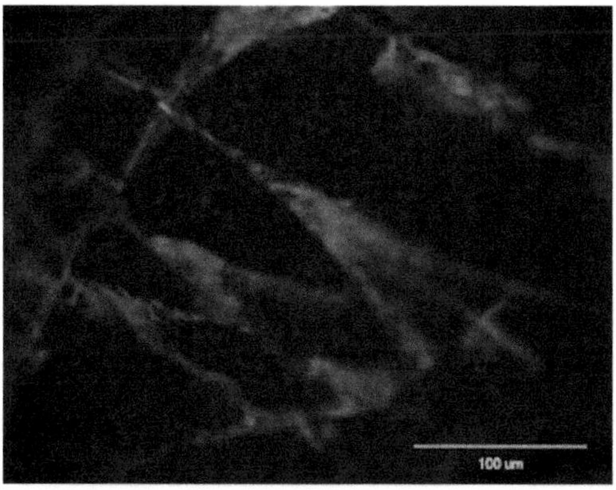

Figure 2.3 Fluorescent cracks in polymers cross-linked by tricinnamates. Reprinted from *Sensors and Actuators B: Chemical*, **134**, S.-Y. Cho *et al.*, A fluorescent crack sensor based on cyclobutane-containing cross-linked polymers of tricinnamates, 822–825. Copyright 2008, with permission from Elsevier.[44]

Similarly, Venkataraman *et al.* demonstrated that mechanical force is capable of inducing the *cis*-to-*trans* isomerization of azobenzenes, changing the spectrum (and therefore the color) of the polymer.[43]

Chung *et al.* showed that polymers that were cross-linked by tricinnamates through irradiation undergo retrocycloaddition upon stress reforming the fluorescent tricinammates.[44] Indeed, when cracking the plastic containing these cross-links, fluorescent cracks appeared (Figure 2.3).

Perhaps, the most widely used mechanophore is spiropyran (SP). SP is a colorless to yellowish non-fluorescent molecule. Upon heating, reducing pH, UV irradiation or increased polarity, SP isomerizes to merocyanine (MC), a red, fluorescent molecule.[45] The change in color and fluorescence is a consequence of the different levels of conjugation between SP and MC. In 2009, Moore *et al.* demonstrated that mechanical stress applied across the SP induces its isomerization to MC.[46] Stress applied to the molecule stabilizes the extended MC form, changing the equilibrium state between the two isomers. This mechanophore has been used effectively in different studies to examine the force distribution in materials.[47]

Additional mechanophores used for the detection of mechanochemical scission were developed based on anthracene Diels–Alder (DA) adducts. Originally developed by Bielawski, these non-fluorescent mechanophores produce fluorescent signal under stress by undergoing retro-DA. Bielawski and Makarov *et al.* used this mechanophore to show anti-Hammond effects that occur during polymer mechanochemistry.[48] Boydston *et al.* used the same mechanophore to study mechanochemistry in star-polymers[49] and Moore *et al.* used it to study mechanochemistry at heterointerfaces.[50]

Craig *et al.* have recently shown that similar color and fluorescence effects can be observed using a Coumarin dimer as a mechanophore.[51] Coumarin dimer absorbs only in the far UV, and undergoes photolysis under strong irradiation. Mechanical stress also leads to scission by a retro [2 + 2] cycloaddition to produce Coumarin, which absorbs in the near UV (320 nm) and fluoresces with a maximum at *ca.* 400 nm (Table 2.5).

Table 2.5 Mechanophores that undergo spectral change.

Mechanophore	Change in spectrum	Ref.
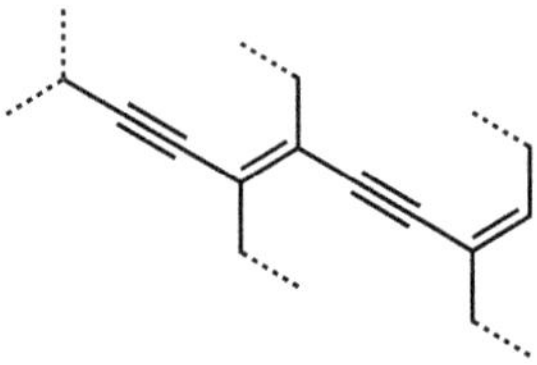	From 615 nm to 475 nm	42
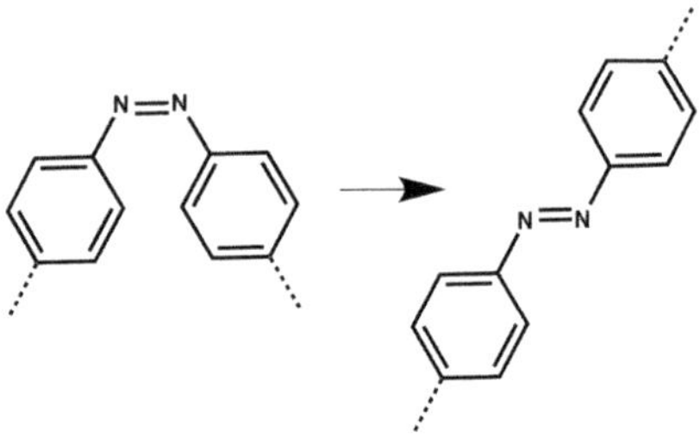	From 440 nm to 337 nm	43
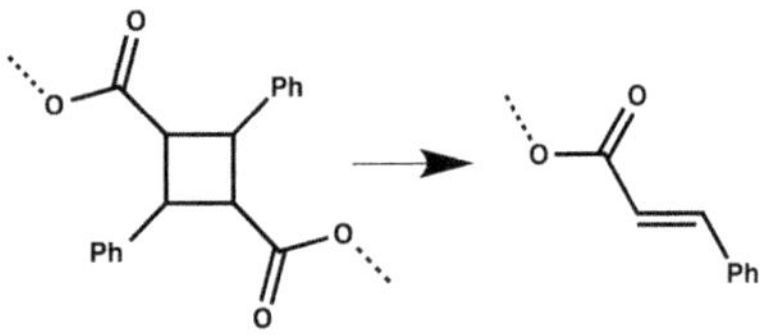	505 nm fluorescence	44
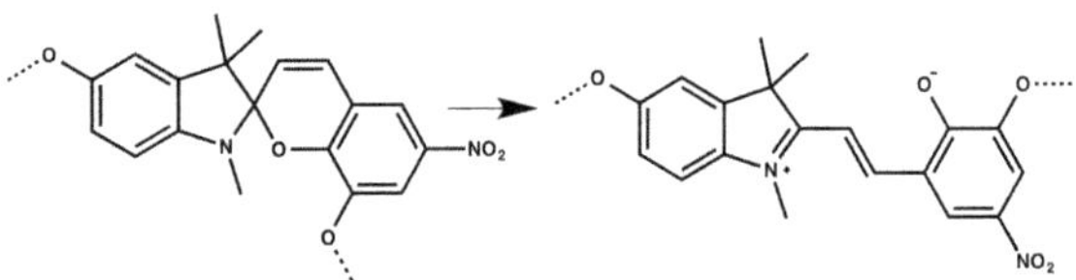	From 400 nm to 560 nm + 620 nm fluorescence	46,47
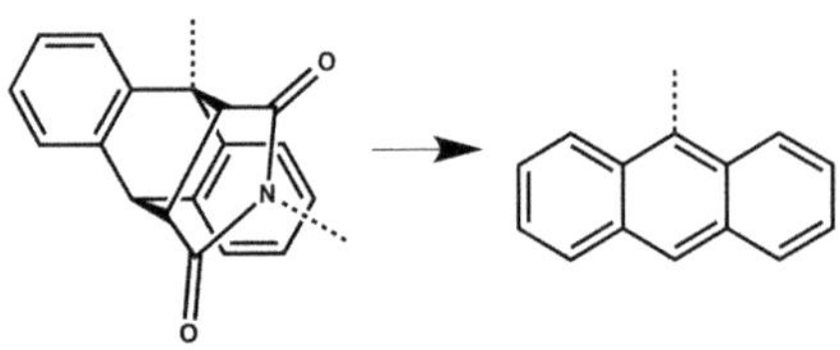	Appearance of 255, 332, 348, 364 and 385 nm peaks	48–50
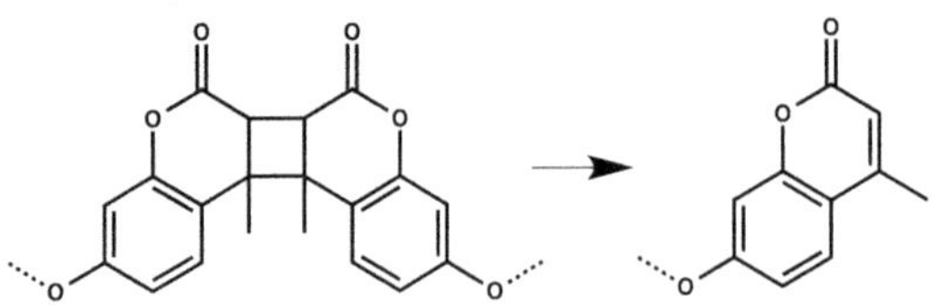	Appearance of 330 nm peak	51

2.5 Production of Small Molecules

In all examples described above, the mechanochemically produced reactant or active catalyst is connected to the polymer network, and while it can easily react in solution, in the solid state it is limited by a very small diffusion. An alternative to the aforementioned examples is to develop mechanophores that produce small molecules that can diffuse faster in the solid state.

Moore *et al.* demonstrated this approach with a mechanophore that produces acid (HCl) under stress.[52] The mechanophore was based on the dihalocyclopropane mechanophores developed previously by Craig *et al.*[15] but, this time, the dichlorocyclopropanation was carried out on a cyclopentadiene instead of an aliphatic double-bond. In this case, following the mechanism demonstrated by Craig, the mechanophore undergoes electrocyclic ring opening to produce a dichlorocyclohexadiene. The latter, in turn, undergoes elimination of HCl at room temperature to form the more stable chlorobenzene. The acid produced has the potential to catalyze bond-forming reactions and thus provide self-reinforcing materials.

Boydston *et al.*[53] and Craig *et al.*[54] worked in parallel on elastomers that could be repeatedly activated and would recover their original shape.[55] In both cases, the mechanophore would undergo a retro DA reaction under stress, releasing a small molecule (Figure 2.4). Instead of pulling the diene and dienophile apart, the new approach used by the authors was to change the bond angles of the DA adduct to destabilize the adduct and push the reaction towards the reactants.

A final example was developed by Moore *et al.* where numerous small molecules are produced per mechanochemical reaction.[56] The authors studied the mechanochemistry of cyclic poly(phthalaldehyde), a polyacetal with low ceiling temperature (T_c). Upon ultrasonication, heterolytic scission of the C–O bond occurs, leading to a zwitterionic intermediate that depolymerizes thermally, since it is above its T_c. Interestingly, after depolymerization, the

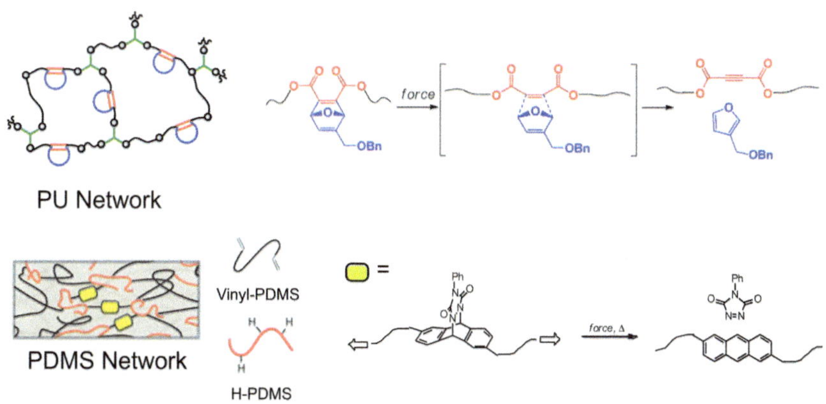

Figure 2.4 Small-molecule releasing mechanophores in elastomers. Reprinted by permission from Macmillan Publishers Ltd: *Nature Chemistry* (ref. 55) Copyright 2014.

mixture of remaining polymer and mechanochemically produced monomer could be cooled below T_c and repolymerized, regenerating the polymer that was lost mechanically. Recently, Goodwin *et al.* showed that alternating polysulfone copolymers also undergo mechanochemical depolymerization to release large amounts of vinyl acetate and SO_2 (Table 2.6).[57]

2.6 Other

Craig *et al.* have shown that dichlorocyclopropanes undergo electrocyclic ring-opening to 2,3-dichloroalkenes, which can be followed by NMR.[58] Recently, he used this mechanochemical reaction as a force probe to measure the mechanochemical stability of different chemical bonds such as C–C,

Table 2.6 Mechanophores that release small molecules.

Mechanophore	Small molecule released	Ref.
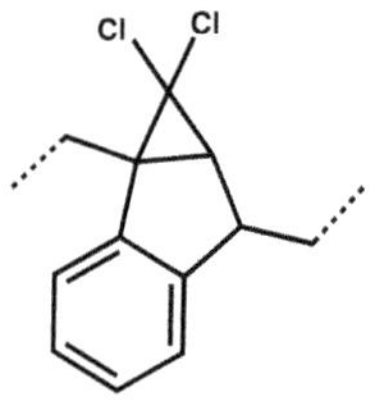	HCl	52
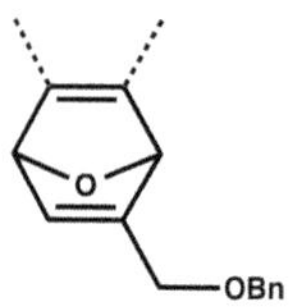	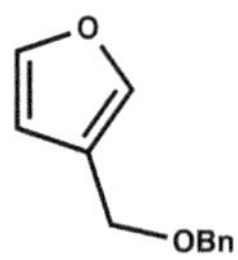	53
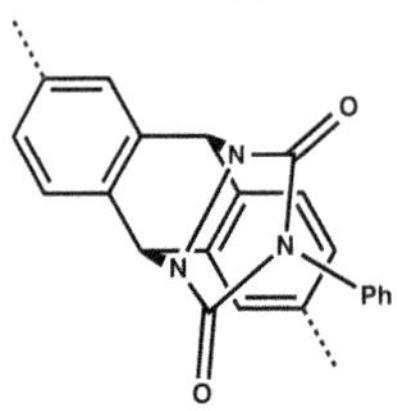	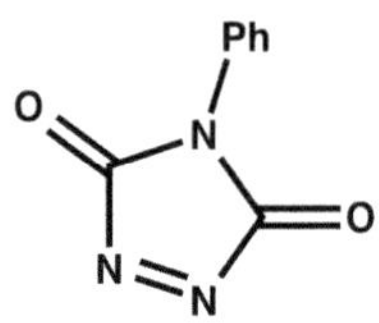	54
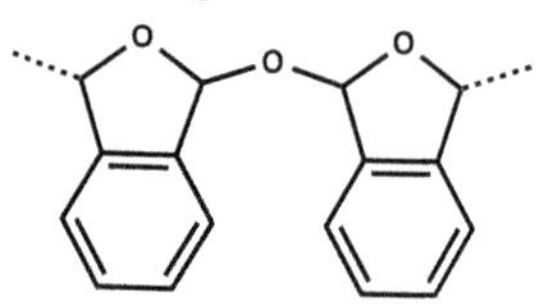	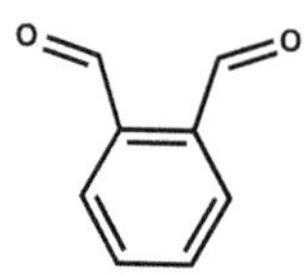	56
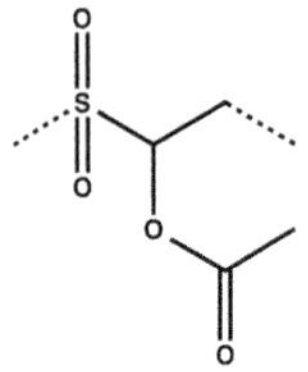	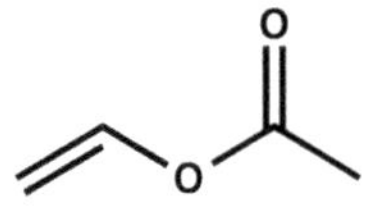 SO_2	57

Table 2.7 Additional mechanophores and their uses.

Mechanophore	Use	Ref.
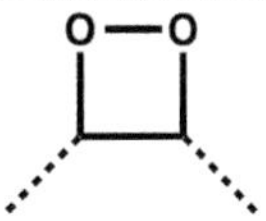	420 nm chemoluminescence	58,59
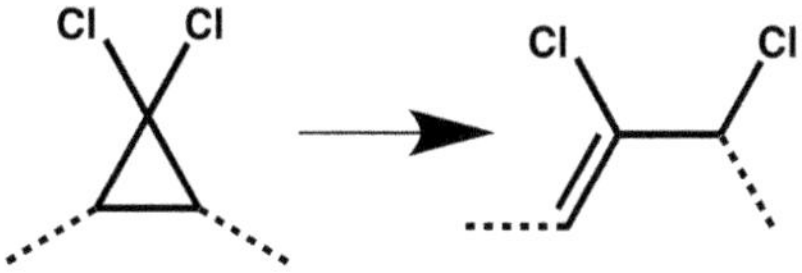	NMR detection of strain release	60,62

C–O and C–S.[59] Importantly, the rate of scission is not directly proportional to bond energy, and the stability of the radical products needs also to be taken into consideration.

Sijbesma *et al.* have used 1,2-dioxetanes as mechanophores.[60] Bis(adamantyl)-dioxetanes were previously shown to be thermally stable in contrast to typical 1,2-dioxetanes.[61] When used as cross-linkers or along the main chain of thermoplastics,[62] they undergo [2 + 2] retrocycloaddition to 2-adamantanone upon application of stress. Importantly, one of the adamantanones is produced in the excited state, and undergoes spontaneous chemoluminescence, emitting a photon around 420 nm. This new mechanophore, like SP, provides an excellent probe to study force distribution in materials (Table 2.7).

2.7 Conclusions and Perspectives

This chapter summarized many mechanophores that were developed with different goals. Many of these mechanophores did not react as originally expected by the authors and their application changed. Looking at the described mechanophores, it is clear which approaches work well in mechanochemistry: bond scissions, retro cycloadditions or electrocyclic ring-openings are repeatedly encountered in the development of mechanochemistry and mechanophores.

Importantly, many of these mechanophores demonstrated their potential in different fields, from catalysts to material science. The scientific knowledge about force distribution in materials and at the molecular level is constantly growing. Yet, in chemistry, there is still a vast open field of mechanically driven reactions to develop. The potential of changing the thermodynamic parameters of reactions, perhaps tuning them by applying different forces, could give access to new reactions and products inaccessible by conventional chemistry. Mechanochemistry has the potential to be a groundbreaking new methodology in synthetic organic chemistry. The chemistry to bind, stress and then release substrates is the next required step to ensure that mechanochemistry remains a prolific field of chemistry.

References

1. M. K. Beyer and H. S. Clausen-Schaumann, *Chem. Rev.*, 2005, **105**, 2921.
2. (a) J. Li, C. Nagamani and J. S. Moore, *Acc. Chem. Res.*, 2015, **48**, 2181; (b) J. N. Brantley, K. M. Wiggins and C. W. Bielawski, *Polym. Int.*, 2013, **62**, 2.
3. R. P. Wool and R. H. Boyd, *J. Appl. Phys.*, 1980, 5116.
4. W. Kauzmann and H. Eyring, *J. Am. Chem. Soc.*, 1940, **62**, 3113.
5. H. Morawetz, *Polymers—The Origins and Growth of a Science*, John Wiley & Sons, New York, 1st edn, 1985, ch. 10, pp. 86–98.
6. H. Staudinger and W. Heuer, *Ber. Dtsch. Chem. Ges.*, 1934, **67**, 1159.
7. J. Sohma, *Prog. Polym. Sci.*, 1989, **14**, 451.
8. M. V. Encina, E. Lissi, M. Sarasúa, L. Gargallo and D. Radic, *J. Polym. Sci., Polym. Lett. Ed.*, 1980, **18**, 757.
9. K. L. Berkowski, S. L. Potisek, C. R. Hickenboth and J. S. Moore, *Macromolecules*, 2005, **38**, 8975.
10. K. Imato, A. Irie, T. Kosuge, T. Ohishi, M. Nishihara, A. Takahara and H. Otsuka, *Angew. Chem., Int. Ed.*, 2015, **54**, 6168.
11. Y. C. Li, A. Nese, K. Matyjaszewski and S. S. Sheiko, *Macromolecules*, 2013, **46**, 7196.
12. W. F. Busse and R. N. Cunningham, *Proc. Rubber Technol. Conf. London*, 1938, 288.
13. R. Groote, B. M. Szyja, F. A. Leibfarth, C. J. Hawker, N. L. Doltsinis and R. P. Sijbesma, *Macromolecules*, 2014, **47**, 1187.
14. J. M. Lenhardt, M. T. Ong, R. Choe, C. R. Evenhuis, T. J. Martinez and S. L. Craig, *Science*, 2010, **329**, 1057.
15. J. M. Lenhardt, J. W. Ogle, M. T. Ong, R. Choe, T. J. Martinez and S. L. Craig, *J. Am. Chem. Soc.*, 2011, **133**, 3222.
16. M. Sakaguchi, M. Makino, T. Ohura and T. Iwata, *J. Phys. Chem. A*, 2012, **116**, 9872.
17. M. Sakaguchi, H. Kinpara, Y. Hori, S. Shimada and H. Kashiwbara, *J. Polym. Sci., Part B: Polym. Phys.*, 1987, **25**, 1431.
18. J. R. Thomas and L. DeVries, *J. Phys. Chem.*, 1959, **63**, 253.
19. T. Shiraki, C. E. Diesendruck and J. S. Moore, *Faraday Discuss.*, 2014, **170**, 385.
20. K. Zhang, X.-H. Xu and F.-L. Qing, *J. Org. Chem.*, 2015, **80**, 7658.
21. C. E. Diesendruck and J. S. Moore, in *Self-healing Polymers: From Principles to Applications*, ed. W. Binder, Wiley-VCH, Weinheim, 2013, ch. 8, pp. 191–211.
22. C. R. Hickenboth, J. S. Moore, S. R. White, N. R. Sottos, J. Baudry and S. R. Wilson, *Nature*, 2007, **446**, 423.
23. J. Wang, I. Piskun and S. L. Craig, *ACS Macro Lett.*, 2015, **4**, 834.
24. M. J. Kryger, M. T. Ong, S. A. Odom, N. R. Sottos, S. R. White, T. J. Martinez and J. S. Moore, *J. Am. Chem. Soc.*, 2010, **132**, 4558.
25. M. J. Kryger, A. M. Munaretto and J. S. Moore, *J. Am. Chem. Soc.*, 2011, **133**, 18992.
26. Z. S. Kean, A. L. Black-Ramirez, Y. Yan and S. L. Craig, *J. Am. Chem. Soc.*, 2012, **134**, 12939.

27. Z. S. Kean, Z. Niu, G. B. Hewage, A. L. Rheingold and S. L. Craig, *J. Am. Chem. Soc.*, 2013, **135**, 13598.
28. M. J. Robb and J. S. Moore, *J. Am. Chem. Soc.*, 2015, **137**, 10946.
29. C. E. Diesendruck, N. R. Sottos, J. S. Moore and S. R. White, *Angew. Chem., Int. Ed.*, 2015, **54**, 10428.
30. H. M. Klukovich, Z. S. Kean, S. T. Iacono and S. L. Craig, *J. Am. Chem. Soc.*, 2011, **133**, 17882.
31. A. L. Black, J. A. Orlicki and S. L. Craig, *J. Mater. Chem.*, 2011, **21**, 8460.
32. A. L. Black-Ramirez, Z. S. Kean, J. A. Orlicki, M. Champhekar, S. M. Elsakr, W. E. Krause and S. L. Craig, *Nat. Chem.*, 2013, **5**, 757.
33. H. M. Klukovich, Z. S. Kean, A. L. Black-Ramirez, J. M. Lenhardt, J. Lin, X. Hu and S. L. Craig, *J. Am. Chem. Soc.*, 2012, **134**, 9577.
34. H. M. Klukovich, T. B. Kouznetsova, Z. S. Kean, J. M. Lenhardt and S. L. Craig, *Nat. Chem.*, 2013, **5**, 110.
35. C. E. Diesendruck, L. Zhu and J. S. Moore, *Chem. Commun.*, 2014, **50**, 13235.
36. R. Groote, R. T. M. Jakobs and R. P. Sijbesma, *Polym. Chem.*, 2013, **4**, 4846.
37. J. M. J. Paulusse and R. P. Sijbesma, *Angew. Chem., Int. Ed.*, 2004, **43**, 4460.
38. S. Karthikeyan, S. L. Potisek, A. Piermattei and R. P. Sijbesma, *J. Am. Chem. Soc.*, 2008, **130**, 14968.
39. A. Piermattei, S. Karthikeyan and R. P. Sijbesma, *Nat. Chem.*, 2009, **1**, 133.
40. R. T. M. Jakobs and R. P. Sijbesma, *Organometallics*, 2012, **31**, 2476.
41. R. Groote, L. van Haandel and R. P. Sijbesma, *J. Polym. Sci., Part A: Polym. Chem.*, 2012, **50**, 4929.
42. M. F. Rubner, *Macromolecules*, 1986, **19**, 2129.
43. S. K. Surampudi, H. R. Patel, G. Nagarjuna and D. Venkataraman, *Chem. Commun.*, 2013, **49**, 7519.
44. S.-Y. Cho, J. –G. Kim and C.-M. Chung, *Sens. Actuators, B*, 2008, **134**, 822.
45. R. Klajn, *Chem. Soc. Rev.*, 2014, **43**, 148.
46. D. A. Davis, A. Hamilton, J. Yang, L. D. Cremar, D. Van Gough, S. L. Potisek, M. T. Ong, P. V. Braun, T. J. Martínez, S. R. White, J. S. Moore and N. R. Sottos, *Nature*, 2009, **459**, 68.
47. (a) C. K. Lee, D. A. Davis, S. R. White, J. S. Moore, N. R. Sottos and P. V. Braun, *J. Am. Chem. Soc.*, 2010, **132**, 16107; (b) C. M. Kingsbury, P. A. May, D. A. Douglas, S. R. White, J. S. Moore and N. R. Sottos, *J. Mater. Chem.*, 2011, **21**, 8381; (c) B. A. Beiermann, D. A. Davis, S. L. B. Kramer, J. S. Moore, N. R. Sottos and S. R. White, *J. Mater. Chem.*, 2011, **21**, 8443; (d) C. K. Lee, C. E. Diesendruck, E. Lu, A. N. Pickett, P. A. May, J. S. Moore and P. V. Braun, *Macromolecules*, 2014, **47**, 2690; (e) A.-D. N. Celestine, B. A. Beiermann, P. A. May, J. S. Moore, N. R. Sottos and S. R. White, *Polymer*, 2014, **55**, 4164.
48. S. S. M. Konda, J. N. Brantley, B. T. Varghese, K. M. Wiggins, C. W. Bielawski and D. E. Makarov, *J. Am. Chem. Soc.*, 2013, **135**, 12722.
49. D. C. Church, G. I. Peterson and A. J. Boydston, *ACS Macro Lett.*, 2014, **3**, 648.

50. J. Li, T. Shiraki, B. Hu, R. A. E. Wright, B. Zhao and J. S. Moore, *J. Am. Chem. Soc.*, 2014, **136**, 15925.
51. Z. S. Kean, G. R. Gossweiler, T. B. Kouznetsova, G. B. Hewage and S. L. Craig, *Chem. Commun.*, 2015, **51**, 9157.
52. C. E. Diesendruck, B. D. Steinberg, N. Sugai, M. N. Silberstein, N. R. Sottos, S. R. White, P. V. Braun and J. S. Moore, *J. Am. Chem. Soc.*, 2012, **134**, 12446.
53. M. B. Larsen and A. J. Boydston, *J. Am. Chem. Soc.*, 2014, **136**, 1276.
54. G. R. Gossweiler, G. B. Hewage, G. Soriano, Q. Wang, G. W. Welshofer, X. Zhao and S. L. Craig, *ACS Macro Lett.*, 2014, **3**, 216.
55. N. R. Sottos, *Nat. Chem.*, 2014, **6**, 381.
56. C. E. Diesendruck, G. I. Peterson, H. J. Kulik, J. A. Kaitz, B. D. Mar, P. A. May, S. R. White, T. J. Martinez, A. J. Boydston and J. S. Moore, *Nat. Chem.*, 2014, **6**, 623.
57. K. Kumar and A. P. Goodwin, *ACS Macro Lett.*, 2015, **4**, 907.
58. J. M. Lenhardt, A. L. Black and S. L. Craig, *J. Am. Chem. Soc.*, 2009, **131**, 10818.
59. B. Lee, Z. Niu, J. Wang, C. Slebodnick and S. L. Craig, *J. Am. Chem. Soc.*, 2015, **137**, 10826.
60. Y. Chen, A. J. H. Spiering, S. Karthikeyan, G. W. M. Peters, E. W. Meijer and R. P. Sijbesma, *Nat. Chem.*, 2012, **4**, 559.
61. G. B. Schuster, N. J. Turro, H.-C. Steinmetzer, A. P. Schaap, G. Faler, W. Adam and J. C. Liu, *J. Am. Chem. Soc.*, 1975, **97**, 7110.
62. Y. Chen and R. P. Sijbesma, *Macromolecules*, 2014, **47**, 3797.

CHAPTER 3

Optical Sensing of Stress in Polymers

R. GÖSTL, J. M. CLOUGH AND R. P. SIJBESMA*

Technische Universiteit Eindhoven, Molecular Science and Technology, Group Supramolecular Polymer Chemistry, PO Box 513, 5600 MB Eindhoven, The Netherlands
*E-mail: r.p.sijbesma@tue.nl

3.1 Introduction

The measurement of stress and strain at the molecular level is fundamental to understanding the relationship between structure and mechanical properties in materials ranging from biological tissues to engineering polymers. Improved detection (*i.e.* higher sensitivity, higher resolution) of stress and strain may lead to further insights into biological processes and help the development of sensors, artificial muscles and self-healing materials. Simplified detection methods may also support the maintenance of construction materials, where timely detection of structural damage often requires frequent screening of suspicious parts with error-prone, insensitive methods, such as the detection of cracks by eye, or time-consuming methods involving expensive equipment, *e.g.* industrial radiography.[1] Simplification of this process through the development of materials that self-report mechanical stress is a highly relevant endeavour in the field of functional materials chemistry.[2–4] Arguably, the use of optical detection has distinct advantages both in fundamental studies as well as in engineering applications, because

Polymer Chemistry Series No. 26
Mechanochemistry in Materials
Edited by Yoan C. Simon and Stephen L. Craig

Published by the Royal Society of Chemistry, www.rsc.org

light with its high spatiotemporal resolution allows for precise localization of damage and its non-invasive character enables remote visualization. Additionally, light's high energetic resolution renders simultaneous detection of different colours possible.

In this chapter, we will focus on the incorporation of optical feedback pathways for mechanical stress into polymers, as they arguably constitute the most important group of materials for which functional moieties can be tailor-made and modularly incorporated. For other stress-reporting (liquid-)crystals, small molecules or biomacromolecules the reader is referred to an excellent choice of overview articles.[4–11] In the following, we first classify the fundamental physicochemical pathways to generate optical feedback through mechanical stress and will then lead the reader through the different motifs that have been reported employing these pathways.

The fundamental requirement for the visualization of stress is that the polymer changes its optical properties upon the application of force. Most photochemical and -physical phenomena can be found and employed in a mechanochemical context as well (Figure 3.1) and as such one of the most fundamental pathways is the stress-induced change in absorption.[12] This means that light is absorbed by a chromophore that has been generated or altered through mechanical stress. Subsequently, the material appears in the colour complementary to the absorbed light. A change in absorption (and thus mechanical stress) is easy to quantify *via* the Beer–Lambert law if the molar absorptivity ε and the concentration of the corresponding chromophore are known. However, the Beer–Lambert law also expresses the method's intrinsic disadvantage, namely its low sensitivity, as the measurement relies on the transmission of attenuated light, *i.e.* a small signal against a noisy background. Hence, absorption measurements are commonly not suitable for chromophore concentrations below 10^{-5} M without any special techniques, such as the use of photonic materials that induce absorption not

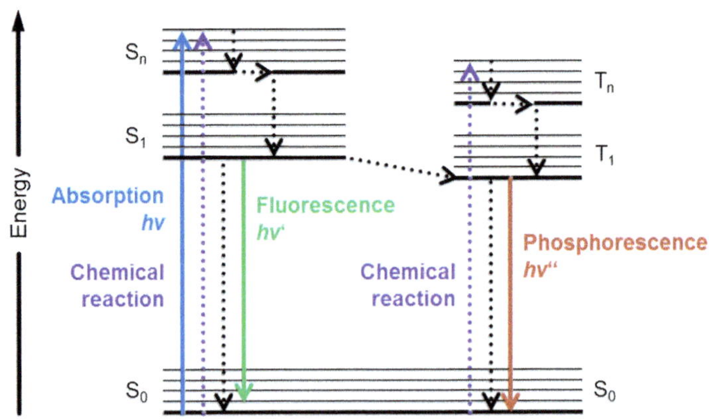

Figure 3.1 Simplified Jablonski diagram illustrating the fundamental transition pathways between ground and excited states.

on a molecular but on an ensemble level.[13] In these materials, the periodic structuring of dielectrics can produce a stopband that prevents light of certain wavelengths from propagating – similar to opals.

Since the absolute intensity of emitted light is easier to detect than the relative intensity of absorbed light, mechanically induced fluorescence can be regarded as the more sensitive optical steady-state detection method.[14] For fluorescence to occur, a photon has to be absorbed by a fluorophore, exciting it into a higher singlet state S_n that converts back to the S_1 state from which a radiative transition can take place, *i.e.* a photon of lower energy is emitted while the fluorophore converts back to the S_0 state. This method is sensitive enough that with well-performing fluorophores, fluorescence is routinely detectable in concentrations from 10^{-7} M, even though measurements of concentrations down to 10^{-12} M have been reported using confocal microscopy.[15] The resulting principal advantage is that fluorogenic compounds can be incorporated at lower concentrations than chromogenic motifs, causing fewer alterations of the original polymer's properties. However, quantification of fluorescence is more difficult than absorption as the efficacy of fluorescence, expressed in its quantum yield Φ_F, is influenced by a number of photophysical events including internal conversion, vibrational relaxation and intersystem crossing to the Triplet state (T), which, in turn, are all dependent on the fluorophore's physicochemical microenvironment.

Mechanically induced chemiluminescence is closely related to fluorescence with the main difference being that the excited states are not populated through the absorption of light but through a preceding chemical reaction. The detectability lies in the same concentration range but since chemiluminophores generally only undergo one chemical reaction before they are consumed, the excited state is not populated continuously but only once for each chemiluminophore rendering chemiluminescence a temporally resolved, non-steady-state method. Similar to super-resolution microscopy techniques,[16,17] chemiluminescence is thus essentially able to achieve a higher spatial resolution beyond the diffraction limit through the recording of successive emission events. Quantification of chemiluminescence has the same difficulties as fluorescence with spin selection rules adding further complication. Through the chemical reaction pathway a part of the excited states created are not only singlet but triplet states with their own associated lifetimes, conversion- and emission-kinetics (phosphorescence).

3.2 Bond–Isomerization Reactions

Isomerization reactions transform molecules by changing the constitution, conformation or configuration of their atoms. Photochromic isomerization reactions[18] were of particular interest to researchers developing polymers with useful mechanical responses. Azobenzene, one of the first photochromic compounds investigated, undergoes a colour-change when isomerizing from the *E*- to *Z*-form and is also affected by mechanical stress because the isomerization is accompanied by a significant geometrical change

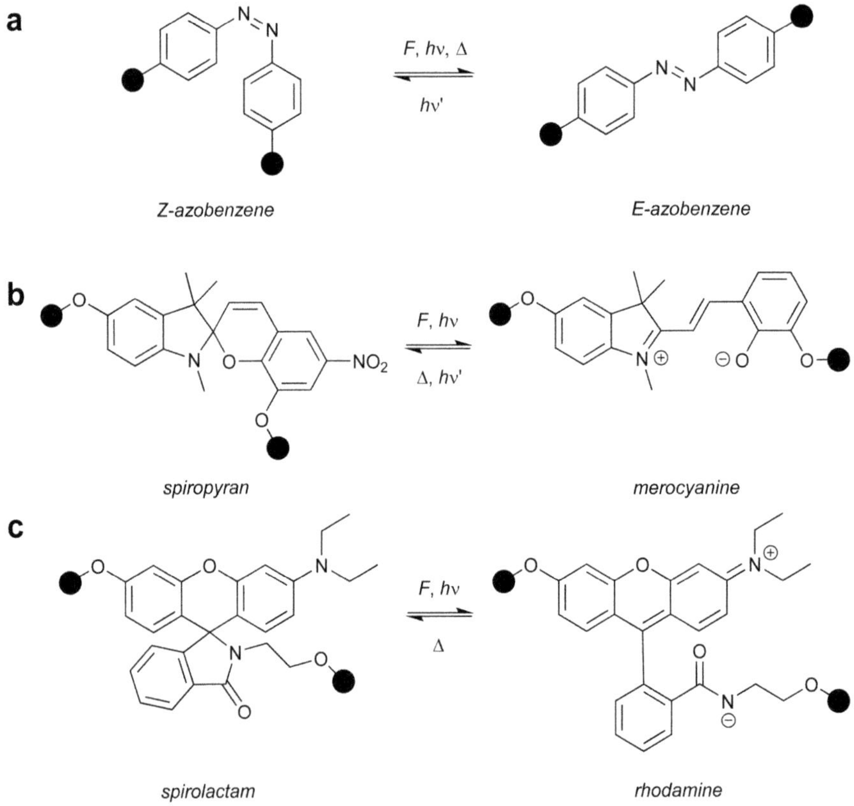

Scheme 3.1 Isomerization reactions induced by force of (a) azobenzene, (b) spiropyran and (c) rhodamine. Black dots represent the attachment point to the polymer architecture.

(Scheme 3.1a).[19–21] Reneker and Kim incorporated the thermodynamically stable *E*-isomer into linear polyurethane chains forming a physically cross-linked network. Light-induced isomerization yielded the unstable *Z*-isomer which could then be transformed to its *E*-form by tensile stress reflected in an altered UV-Vis absorption spectrum. The reversibility of this process renders this the first polymer-based mechanochromic reaction reported. However, the difference between the colours of *E*- and *Z*-isomer is not very pronounced, making azobenzene a non-ideal optical probe for mechanical stress. Interestingly and analogously to the radical initiator azobisisobutyronitrile, ultrasonication of an alkylic azo-moiety incorporated in the centre of linear poly(ethylene glycol) leads to site-specific cleavage of the moiety rather than isomerization.[22]

Arguably, the most successful mechano(fluoro)chromic reaction is the spiropyran to merocyanine isomerization (Scheme 3.1b). First reported by Moore, White, Sottos and coworkers in 2007[23,24] it quickly became the most widely employed optical probe for mechanical stress.[25] Initially experiments

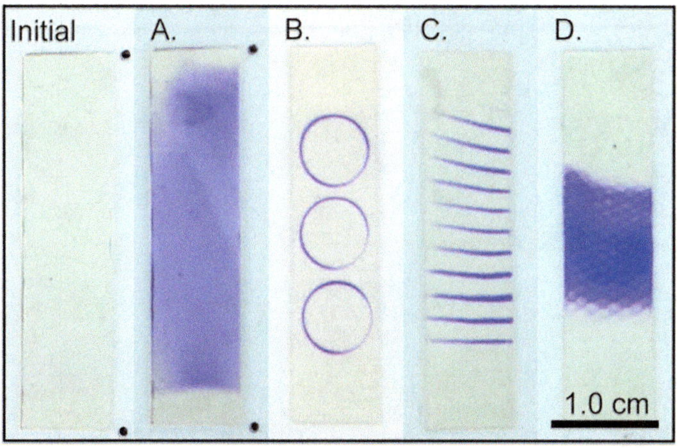

Figure 3.2 Exemplarily shown, localized activation of spiropyran to merocyanine by compression of functionalized poly(dimethylsiloxane) film surfaces. The film can be activated by tension (A) or by rolling a patterned object across the surface to generate circles (B), lines (C) or a cross-hatch pattern (D). Reprinted with permission from G. R. Gossweiler, G. B. Hewage, G. Soriano, Q. Wang, G. W. Welshofer, X. Zhao and S. L. Craig, *ACS Macro Lett.*, 2014, **3**, 216–219.[30] Copyright (2014) American Chemical Society.

showed that incorporated into linear[23] as well as cross-linked[24] poly(methyl acrylate), poly(methyl methacrylate),[26,27] poly(ε-caprolactone),[28,29] poly(dimethylsiloxane),[30] poly(*tert*-butyl acrylate)-*b*-poly(*N*-isopropyl acrylamide)[31] or supramolecular polymers[32–34] the violet-coloured and red-fluorescent merocyanine form could be generated by the application of force to the respective polymers (an example of which is depicted in Figure 3.2).

Furthermore, as soon as the potential of this optical probe was discovered fundamental questions over the effect of stress on mechanical bond activation, such as those of strain hardening and flow,[35] microphase separation,[36] local temperature,[37] crystallinity[38,39] and plasticity/elasticity,[30,39–41] could be tackled leading to valuable insights into the isomerization mechanism and fundamental design principles of mechanophores. Although mechanophore activation is generally achieved in the plastic deformation regime, Craig and coworkers demonstrated that spiropyran can also be isomerized in the elastic deformation regime of poly(dimethylsiloxane).[30] Moreover, Craig and coworkers determined the force necessary to perform the spiropyran to merocyanine isomerization to be as little as ~240 pN by pulling a single polymer chain with an AFM tip.[42] This value lies well below the force necessary to cleave other common covalent mechanophores, retrospectively explaining the widespread success of this excellent mechanophore.

Analogously to spiropyran, Jia and coworkers discovered that various spirolactam derivatives can switch "on" their fluorescence upon force-induced transformation to their rhodamine forms in the solid[43,44] as well as in the

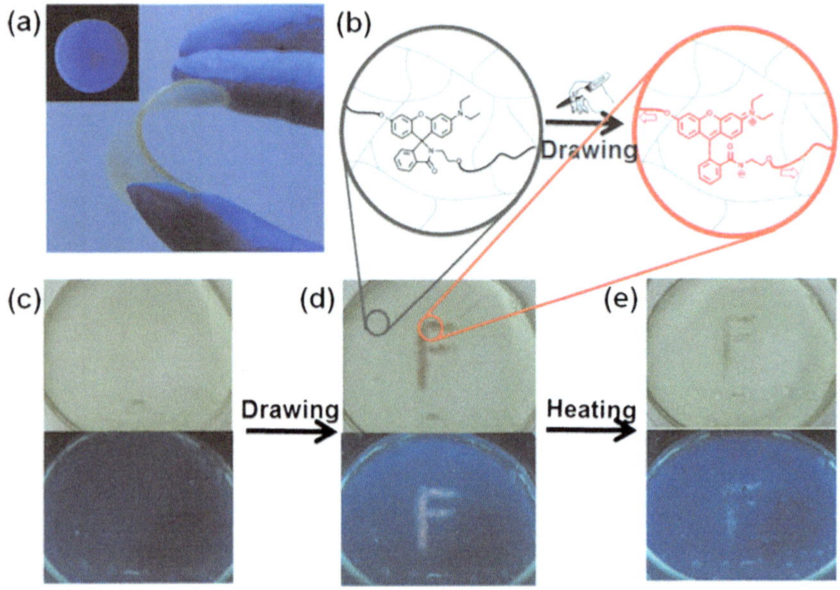

Figure 3.3 Spirolactam mechanophore incorporated into a polyurethane film as (a) macroscopic photography and (b) the chemical structure change to rhodamine after drawing. (c) Unaltered film, (d) drawn on film and (e) heated film under daylight as well as UV-light. Reprinted from ref. 46 with permission from John Wiley and Sons. Copyright © 2015 WILEY-VCH Verlag GmbH & Co. KGaA, Weinheim.

gelated state (Scheme 3.1c).[45] In 2015, they found that incorporation of this motif into a chemically cross-linked polyurethane network and transmission of the force *via* the polymeric backbone facilitates this reaction so that fluorescence can be activated simply by drawing on the polymer film (Figure 3.3).[46] The ring-opening is reversible and thus heating the network restores the initial, non-fluorescent spirolactam form.

The multiple examples given above illustrate the benefit of employing isomerization reactions for the optical sensing of stress, *i.e.* their reversibility rendering the motifs mechano(fluoro)chromic. In contrast to mechanically induced scission reactions (*vide infra*), forward and backward reactions of isomerizations follow a unimolecular rate law. Hence, the mechanophores can in principle be activated repeatedly with comparable kinetics at low concentrations and even before the material fails.[30,41,46]

3.3 Covalent Bond Scission Reactions

Covalent bond scission reactions were the first mechanically induced reactions observed in polymers and reported by Staudinger and Heuer already in 1934.[47] However, they were utilized only recently to generate optical feedback to signal scission events.[48] Cycloelimination reactions are widely exploited bond scission reactions for this purpose. The well-investigated retro Diels–Alder reaction of the anthracene-maleimide adduct is particularly well suited for this purpose as it allows the transition from the optically transparent Diels–Alder adduct with interrupted π-conjugation towards the fully conjugated and fluorescent anthracene *via* the retro Diels–Alder reaction (Scheme 3.2a).

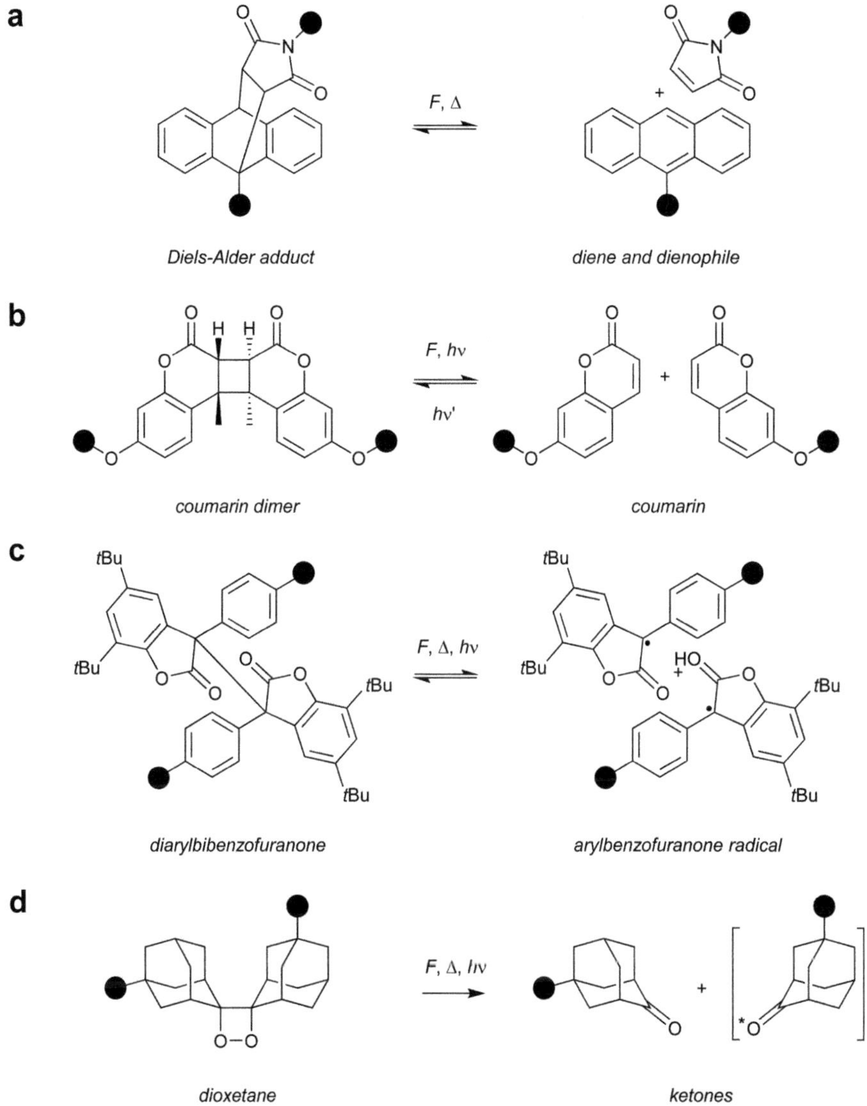

Scheme 3.2 Scissile reactions induced by force of (a) Diels–Alder adduct of anthracene and maleimide, (b) coumarin dimer, (c) diarylbibenzofuranone and (d) dioxetane. Black dots represent the attachment point to the polymer architecture.

Yoshie and coworkers were the first to use this motif in 2011 in a self-mendable polymer network consisting of anthracene-terminated short polyester chains reversibly cross-linked with a trifunctional maleimide unit.[49] Tensile stress could activate the retro Diels–Alder reaction thus releasing anthracene upon bond scission, which was visualized by UV-Vis spectroscopy. However, as many anthracene units in the network's backbone are inevitably unreacted in this system, the stress-sensing is relatively insensitive because it depends on small variations in absorption. Makarov, Bielawski and coworkers incorporated the same Diels–Alder adduct in the centre of a linear poly(methyl acrylate) chain and induced the cleaving event by ultrasonication.[50] As no free anthracene was present in the original polymer, the scission could be quantified by absorption spectroscopy more reliably. Moore and coworkers used this Diels–Alder adduct as an anchoring motif and also followed the ultrasound-induced scission of polymer hairs attached to silica nanoparticles *via* the changes in absorption.[51] Anthracene's fluorescence was first exploited for the detection of bond rupture by Boydston and coworkers, investigating the scission behaviour of star-shaped polymers,[52] as well as by Craig and coworkers, inducing the release of triazolinedione (the most reactive dienophile known)[53] by planarization of the corresponding Diels–Alder adduct cross-linker in a poly(dimethylsiloxane) network.[30] Very recently, Sijbesma and Göstl enhanced the sensing capability of this motif by π-extension of the anthracene unit generating a fluorophore with high Φ_F as well as red-shifted absorption and emission spectra.[54] Hence, by employing the corresponding maleimide Diels–Alder adduct as a cross-linker in a poly(hexyl methacrylate) network mechanical stress was made visible by eye after the release of the fluorophore (Figure 3.4). The fluorescence of anthracene was also used to sense microcracks not *via* the retro Diels–Alder reaction but by cycloelimination of anthracene dimer cross-linkers in a poly(vinyl alcohol) network.[55]

Another force-induced chromo- and fluorogenic cycloelimination reaction was discovered recently by Craig and coworkers.[56] Similar to other

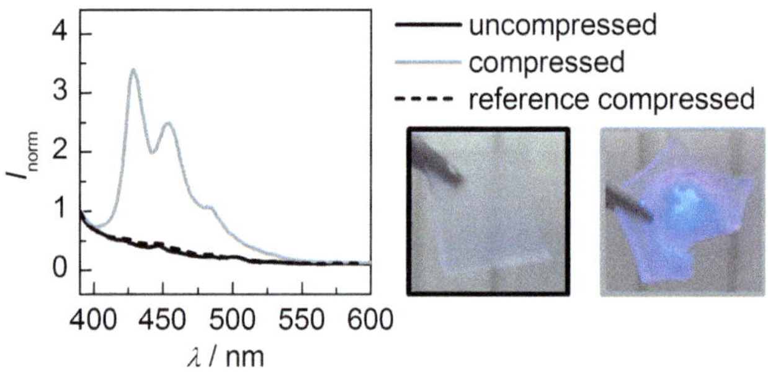

Figure 3.4 Force-induced retro Diels–Alder reaction in a poly(hexyl methacrylate) network generating a π-extended anthracene as visualized by solid-state fluorescence spectroscopy. The boxes show corresponding pictures of the samples while irradiated with a UV hand lamp. Adapted from ref. 54 with permission from the Royal Society of Chemistry.

cyclobutane mechanophores,[57,58] ultrasonication of a widely transparent coumarin dimer incorporated in the centre of linear poly(methyl acrylate) could induce the [2 + 2] cycloelimination and generate fluorescent coumarin chain ends (Scheme 3.2b).

Otsuka and coworkers followed an approach pioneered by Löwenbein and coworkers[59–62] already in the 1920s by incorporation of chromogenic, homolytically cleaving diarylbibenzofuranones as cross-linkers into polymer networks (Scheme 3.2c). However, due to rapid thermal radical recombination visualization of their bright blue colour is demanding and was initially only possible by exerting force through freezing of organic solvent-swollen networks and thermal trapping of the radical.[63] Nevertheless, incorporation of a high concentration of this mechanophore in a physically cross-linked dynamic covalent polyurethane network led to the characteristic blue colour through the application of strain.[64]

Alongside the efforts to develop force-induced chromo- or fluorogenic reactions, the Sijbesma group has established mechanically induced chemiluminescence (mechanoluminescence) as an excellent method to detect strain in polymeric materials. In this strategy, thermally stable bis(adamantyl)-1,2-dioxetane is incorporated in a polymer chain or network and upon the application of force, the dioxetane group decomposes *via* a cycloelimination to yield two ketones under the emission of light (Scheme 3.2d). This phenomenon was first demonstrated in 2012 from dioxetane-functionalized poly(methyl acrylate), both in bulk samples placed under tension and in solutions of linear polymers upon sonication.[65] The advantages bestowed by the transient nature of mechanoluminescence were evident even in this first study, allowing both the location and temporal progression of bond scission to be visualized with high sensitivity.

Employing this work as foundation, dioxetane mechanoluminescence shed light on two new approaches to toughening polymer networks, both of which enhance a material's ability to dissipate strain energy before fracture. The first strategy, devised by the Creton group, entails the incorporation of second and third networks in the first elastomer network *via* sequential free radical polymerizations, forming a multiple network elastomer. In single edge notch tests on dioxetane-functionalized multiple networks, light emission was highly localized at the crack tip for the single network, becoming more intense for the double network and for the triple network, a large yielding zone could be visualized over an extended region ahead of the crack tip (Figure 3.5).[66] Mechanoluminescence thus confirmed the significance of the first and second networks as sacrificial stress-bearers, dissipating energy prior to material failure. Furthermore, the mechanoluminescence traces were rich in information about the yielding zone, revealing its extent and shape, which could be controlled by the extensibility of the second network chains.

A second reinforcement strategy, investigated by the groups of Craig and Sijbesma, exploited the stress-homogenizing effect of weak interactions between pendant pyridines and dissolved palladium pincer complexes.

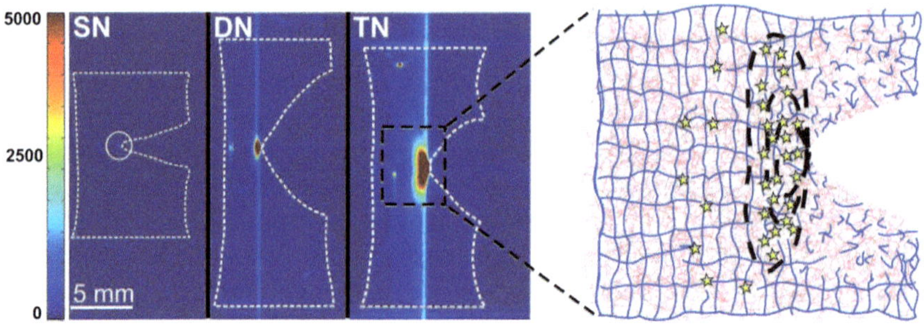

Figure 3.5 Dioxetane mechanoluminescence upon crack propagation from a notched edge in samples comprising a single network (SN), a double network (DN) and a triple network (TN). From ref. 66. Reprinted with permission from AAAS.

In such a system, the supramolecular cross-links have little effect on the modulus and structure of the material, but boost the fracture energy. The dioxetane mechanophore demonstrated that covalent bond scission was indeed inhibited by the addition of reversible cross-linkers to the network, with higher concentrations of supramolecular cross-linkers delaying the onset of mechanoluminescence to a greater extent.[67]

The mechanisms of deformation exhibited by common engineering polymers have also been probed with mechanoluminescence. Sijbesma and Chen addressed the role of microphase separation into soft and hard domains in the deformation behaviour of polyurethanes.[68] The dioxetane mechanophore was incorporated in segmented copolymers comprising poly(tetramethylene oxide) (PTMO) soft segments with different hydrogen-bonding hard segments. Stronger hydrogen-bonding in the hard segments inhibited chain disentanglement and the ability of the material to accommodate the applied strain, leading to increased mechanoluminescence emission. In addition, increasing molecular weight increased the number of hard blocks per chain, reducing chain slippage and increasing mechanoluminescence emission.

Complementary to efforts to establish the general utility of this mechanophore, the excited state products of the mechanical scission process were characterized upon sonication of linear PMA polymers and found to consist of singlet and triplet excited ketones in a 1 : 10 ratio, with an overall chemiexcitation yield of approximately 10%.[69] Given that the decomposition of alkyl dioxetanes generally gives an unusually large excess of triplet excited states,[70] such information is useful as singlet and triplet excited ketones require different acceptors to maximize luminescence efficiency.

Very recently, the Sijbesma group presented a second system for mechanically induced chemiluminescence and combined mechanically activated bis-*N*-heterocyclic carbene (NHC) metal complexes with two latent chemiluminescent substrates bearing acidic protons: a 2-coumaranone derivative and the well-known spiroadamantyl-substituted phenolic 1,2-dioxetane.

Upon sonication of the polymeric bis-NHC complexes in dilute solution with one of the two chemiluminescent substrates, mechanically released NHCs deprotonated the substrates, activating the emission of light.[71] In addition to the higher chemiluminescence efficiency exhibited by these luminophores, there is future scope to tune the force threshold at which mechanoluminescence is obtained by varying the structure of the complex. It is hoped too that this process could be made catalytic, as opposed to stoichiometric.

Clearly, the most sensitive optical probes reported can be found amongst the covalent scission reactions. However, although most of the described reactions are principally reversible, in contrast to the mechano(fluoro)chromic reactions described in Section 3.2 the bimolecular kinetics often render the back reaction factually impossible at low concentrations.

3.4 Conjugated Polymers

Conjugated polymers can exhibit mechanically induced fluorochromism and fluorochromogenic transitions, with poly(diacetylenes) being the archetypal example in this area. These polymers, formed from the topochemical polymerization of supramolecularly assembled diacetylene monomers, possess an alternating ene-yne backbone structure. The resulting π-delocalization along the backbone coupled with the conformational restriction provided by the packing of the side-groups gives rise to these materials' dramatic optical responses to a wide variety of stimuli, including heat, solvent, force, magnetic fields, current, ions and ligand binding.[72–74] A strong optical absorption from the π-π* transition of the backbone is observed at *ca.* 640 nm, which is generally blue-shifted to *ca.* 540 nm following a chromic or chromogenic transition, causing the material's colour to change from blue to red or yellow. Furthermore, in the activated red-state the material is strongly fluorescent, whereas in the starting blue-state fluorescence is symmetry-forbidden. It is thought that the backbone of the polymer is formed in a strained state during polymerization and the stimulated optical changes arise from the relaxation of the backbone into a lower energy conformation *via* rotation about the C–C bonds, altering the planarity of the backbone and hence the degree of π-orbital overlap. Theoretical calculations indicate that rotation of only a few degrees is necessary to give rise to a significant blue-shift in the absorption band, making PDAs sensitive to small changes in the strain experienced by its backbone.

The thermally induced chromic and chromogenic responses of PDA have been the most deeply investigated, although the use of mechanical force as a stimulus for these materials has received more attention in recent years. In 1991, Nallicheri and Rubner were the first to use PDA as a force sensor in a polymeric material.[75] They synthesized segmented polyurethanes containing diacetylenes in their hard blocks which could be polymerized in the solid state; upon the application of tensile stress, the poly(diacetylene)s underwent a chromic phase transition which was reversible up to 350%, in addition to aligning perpendicularly to the direction of tensile strain.

Similarly, Sijbesma and coworkers prepared copoly(ether urea) thermoplastic elastomers containing polymerizable diacetylene units in their hard segments, buttressed on both sides by hydrogen-bonding urea groups.[76] Upon irradiation with UV light the diacetylene units formed highly ordered PDAs, although the chromic transition in this case was reversible only up to 80%. The difference between these two materials was ascribed to a lower strain onset for hard block disruption in the poly(ether urea).

More recently, research has shifted towards more specific technological applications. For example, Tao, Li and coworkers reported in 2013 on the chromic transition of PDA in nanowires of poly(ethylene oxide) (PEO) induced by the mechanical force from an oscillating tuning fork.[77] In 2014, Kim and coworkers combined the selective swelling properties of PDMS with the mechanical sensitivity of poly(diacetylene) to create colorimetric sensors for aliphatic hydrocarbons based on PDA-functionalized PDMS networks.[78] Outside of the realm of bulk polymers, mechanochromism from PDAs has been probed on the nanoscale in two-dimensional PDA monolayers, both by varying the lateral surface pressure of a monolayer in a Langmuir–Blodgett trough[79] and by scratching a monolayer surface with a scanning force microscopy tip.[80,81] Under both forms of mechanical activation, the red-phase of PDA was obtained.

Other conjugated polymers also exhibit strain-dependent conformational changes, most notably poly(fluorenes).[82,83] One of the latest innovations in this area was the report by Sprakel and coworkers in 2015 describing a π-conjugated poly(fluorene)-based polyelectrolyte as a strain sensor for protein complexation.[84] The unbound anionic poly(fluorene) rotates relatively freely in solution. Upon binding to neutral-cationic coat proteins, the polymer stretches and the fluorene groups are brought into plane, lengthening the effective conjugation length and leading to marked changes to the vibronic peaks in the fluorescence spectrum. The mechanical sensing capability of this probe depends on the backbone adopting a more strained conformation when activated, in contrast to the PDA-based sensors, which are thought to release their conformational strain in the activated red-phase.

Moreover, Swager and Kim showed in 2001 that surfactant poly(*p*-phenylene-ethynylene) Langmuir films can change their colour reversibly upon the application of mechanical stress due to conformational rearrangement of their conjugated backbone.[85]

3.5 Chromophore Rearrangement

Alongside approaches employing covalent incorporation of mechanophores, physically incorporated chromophores can also be used to imbue polymer blends with optical mechanoresponsivity. Pigmented dyes have been added to polymeric materials since the dawn of the polymer era in the 1930s, but the application of this to the sensing of force has only been explored quite recently.[3,86,87] The so-called "aggregachromic" (in most cases more correctly aggregachromogenic) dyes are generally flat, rigid, aromatic molecules, with electron-releasing or electron-withdrawing groups conjugated to the π-system, promoting their self-assembly *via* π-π and electrostatic interactions into micro- or nano-sized aggregates during processing. The application of force disrupts these supramolecular interactions, breaking up the aggregates and leading to significant optical changes in the material. Generally, this method is easier to implement than the covalent approaches outlined above, requiring only that the polymer is semi-crystalline and melt- or solvent-processable to blend in the dyes, thus lending itself naturally to studies of industrial polymers on a larger scale. However, the disruption of the aggregates occurs in the plastic regime of deformation and is therefore usually irreversible, meaning that the optical changes are (fluoro)chromogenic.

One of the first and most successful aggregachromic dyes were the oligo(phenylene vinylene)s (OPVs), pioneered by the Weder group (Figure 3.6). On dispersing OPVs bearing cyano groups in conjugation with the aromatic system in linear low density polyethylene (LLDPE), the dye molecules formed excimers above a certain concentration of dye, giving rise to a broad excimer emission band at *ca.* 644 nm in addition to the monomer emission peaks at 491 nm and 536 nm.[88] The application of tensile strain to these polymer blends greatly reduced the intensity of the excimer band, indicating break-up of the stacks and molecular dispersion of the monomeric dyes. A melt-processed

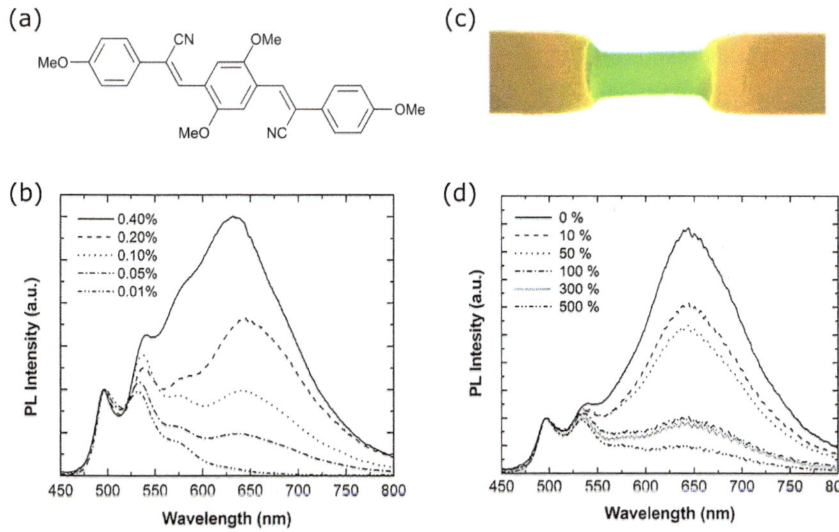

Figure 3.6 The aggregachromic dye, 1,4-bis(R-cyano-4-methoxystyryl)-2,5-dimethoxybenzene (BCMDB) (a) exhibits emission spectra (b) when incorporated in LLDPE at varying concentrations. (c) Fluorescence colour change on straining a 0.18 wt% BCMDB/LLDPE blend to 500% (under excitation with UV light) and (d) the emission spectra of a 0.20 wt% blend on increasing applied strain. Adapted with permission from B. R. Crenshaw and C. Weder, *Chem. Mater.*, 2003, **15**, 4717–4724.[89] Copyright (2003) American Chemical Society.

blend of LLDPE and 0.2 wt% 1,4-bis(α-cyano-4-methoxystyryl)-2,5-dimethoxybenzene (BCMDB) exhibited measurable excimer disruption at strains as low as 10% and by 500%, the OPV molecules were almost completely dispersed within the matrix.[89]

By varying the structure of the dye, the composition of the blend and the processing conditions, the phase behaviour of the blends could be controlled to obtain higher levels of mechanical activation.[89–91] In semi-crystalline PE, smaller aggregates and host matrices with a greater crystalline content were both found to promote aggregate disruption and give a greater mechanical response. Smaller aggregates could be obtained through the use of OPVs with solubilizing alkyl tails, which nucleate more slowly, and through rapid quenching following melt-processing and subsequent solvent plasticization. From a mechanistic viewpoint, dye aggregate rupture is thought to be encouraged by plastic deformation processes such as macromolecular chain pull-out from the crystalline lamellae and the formation of microfibrils: smaller aggregates are easier to break up, whilst a higher crystalline content provides more opportunities for aggregate rupture. Following these initial investigations on PE host matrices, the scope of OPV-based dyes was extended to other polymeric materials, including poly(vinylidene fluorides)[92] and polyesters.[93,94] Similar conclusions concerning the requirements for mechanical activation were drawn in these investigations, demonstrating the broader utility of this approach. However, blended OPV/polyurethanes demonstrated surprisingly little mechanical response, which was attributed to macrophase separation; to overcome this, OPV was covalently incorporated in the polyurethane and greater emission colour changes were then obtained.[95]

The success of the OPVs inspired others to design new aggregachromic dyes. Foremost amongst these are bis(benzoxaloyl)stilbene (BBS) and the perylene derivatives, both of which were first developed by the Pucci group. The mechanically activated fluorochromogenic transitions of these dyes work on a similar principle to the OPVs, in that force induces rupture of supramolecular aggregates of dye molecules, giving an overall blue-shift in the fluorescence emission. The characteristic mechanofluorochromogenic response from BBS, from green excimer fluorescence to the blue fluorescence of the molecularly dispersed monomer, has been demonstrated in poly(propylene) (PP),[96] poly(ethylene),[97] poly(1,4-butylene succinate) (PBS)[98] and polyurethanes.[99–101] Perylene has been blended with PE[102] and upon mechanical activation, the colour of the fluorescence emission changed from yellow to blue, whilst the fluorescence emission of perylene bis-imides changed from orange-red to green-yellow on mechanical activation in PE[103] and PVA.[104] The mechanofluorogenic naphthalimide-coumarin dyad[105] as well as the Cy3-Cy5 pair[106] could be used to modulate fluorescence resonance energy transfer mechanochemically. Organometallic complexes have also been dispersed in polymeric materials to obtain mechanochromism, such as the square-planar platinum(II) complexes of the 4-dodecyloxy-2,6-bis(*N*-methylbenzimidazol-20-yl) pyridine ligand, which was blended in a series of methacrylate polymers.[107]

Moreover, Simon, Weder and coworkers employed microencapsulated hexamethylbenzene and chloranil in poly(dimethylsiloxane) that form coloured charge transfer complexes when the polymer is deformed.[108]

Lastly, dyes that display aggregation induced emission (AIE) have been the focus of an intense research effort since the early 2000s and are just starting to be applied as force sensors in polymers.[109] In the free state, dyes in this class undergo intramolecular motions that inhibit their fluorescence, but in the aggregated state, these motions are restricted, allowing fluorescence to occur. They exhibit reversible "on"/"off" switching behaviour in response to a wide variety of stimuli and their incorporation in polymers has been widely reported, although chromisms induced by other stimuli have attracted the most attention up till now.[110] The first mechanochromic AIE-based polymeric material was described by Yoon and coworkers who made PMMA films loaded with an AIE cyanostilbene derivative (20 wt%).[111] Whilst this dye was only "off" in solution, it nevertheless exhibited two phases in the solid state: a meta-stable green-emitting phase and a more thermodynamically stable blue-emitting phase. The blue phase transformed to the green phase on exposure to small shear forces, such as from writing, and exposing to the vapour ensured recovery to the blue phase. "On"/"off" AIE mechanochromism has also been observed from polyurethanes with a small amount of covalently attached tetraphenylethylene, with fluorescence switching "off" at high strains.[112]

3.6 Photonic Polymers and Cholesteric Liquid Crystals

Photonic materials constitute a very particular group of mechanoresponsive, chromogenic materials. In polymeric systems, they are generated either by fixation of photonic crystals through polymerization or directly by self-assembly of block-copolymers bearing blocks of different dielectric properties. Mechanical deformation of the photonic material then changes the optical path length and thus the stopband's exclusion wavelength leading to a colour-change. As the force-induced stopband alteration of photonic materials is covered in detail in the literature,[13,113–116] we will focus on presenting recent highlights only.

While photonic crystals in polymerized arrays have been reported already in the 1990s,[117–119] multi-stimuli responsive photonic materials have been discussed only relatively recently. Ping Gong and coworkers demonstrated that a self-assembled poly(dodecylglyceryl itaconate)-*b*-poly(acrylamide) matrix responded with changes in absorbance not only to stress but simultaneously also to changes in pH and temperature.[120] The same group was also able to develop a soft, durable and rapidly reversible material in which stress could be quantitatively correlated to the shift in the wavelength λ_{max} of the absorption maximum.[121] Chen and coworkers also self-assembled monodisperse magnetic colloid nanocrystal clusters and achieved a similarly

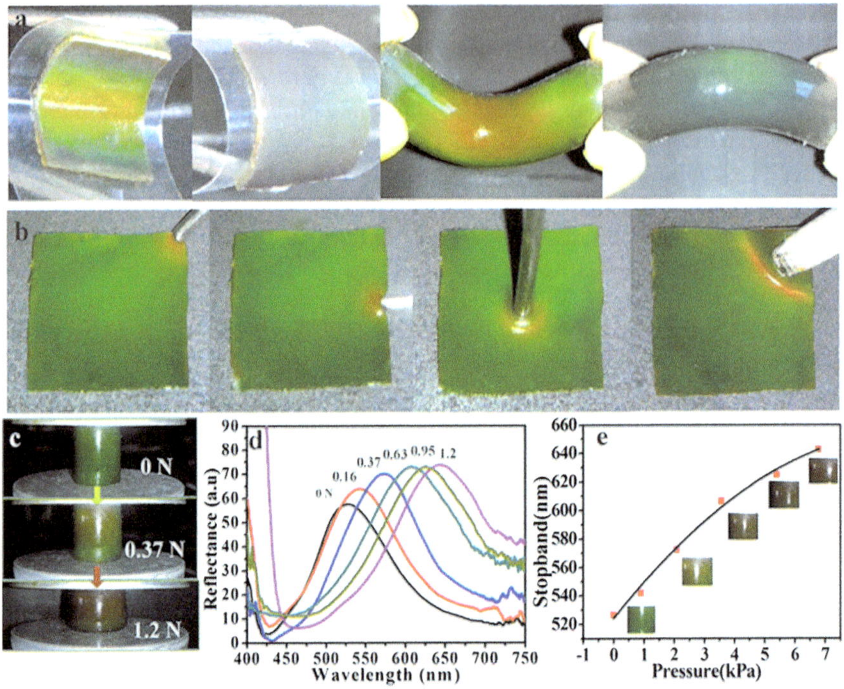

Figure 3.7 Photographs of (a) green photonic films and cylinders under mechanical bending, (b) the same photonic film under multi-directional local deformation, (c) a photonic cylinder under increasing compressive forces accompanied by (d) the diffraction peaks at normal incidence and (e) the corresponding shift of diffraction wavelength as a function of pressure. Reprinted from ref. 122 with permission from John Wiley and Sons. Copyright © 2014 WILEY-VCH Verlag GmbH & Co. KGaA, Weinheim.

durable material, showing a quantitative relationship between stress and shift of λ_{max}.[122] Along with other works,[123,124] these examples pave the way for the implementation of photonic touch sensors in *e.g.* displays (an example of which is depicted in Figure 3.7).

Photonic crystal fibres based on continuous sheets of aligned carbon nanotubes could be produced by Sun, Peng and coworkers and were woven into patterns creating a smart fabric that changed its colour adaptively to deformation.[125] Sheets of photonic crystal paper were also developed by Ge and coworkers which undergo a colour change when the material is relief printed.[126]

Cholesteric liquid crystals (CLCs) respond mechanochromically in a similar way to photonic crystals. CLCs are composed of rod-like molecules (mesogens) arranged in a helical order, with the periodicity described by the pitch of the helix. As a consequence of this structure, circularly polarized light within a particular wavelength range cannot propagate through the matrix and is instead reflected, in effect rendering them 1D photonic band gap materials.[127,128] Changing the pitch of the helix shifts the exclusion wavelength range, giving a colour change. Up till now, transferring mechanical force to the helicoidal structure of the CLC has been achieved predominantly through covalent incorporation of mesogens as chain pendants in a lightly cross-linked elastomeric network. This approach was employed by the Finkelman group in their pioneering development of mechanically tunable lasing materials.[129,130] The mesogens were incorporated in a PDMS matrix, along with a laser dye; the helical order could be induced by anisotropic deswelling of cross-linked samples[129,131] or by orientation between glass plates followed by photo-polymerization.[130] On applying biaxial tension, the wavelength of the reflected laser beam varied in the range of 544–630 nm, caused by a decrease in the pitch with applied stress.[129] Further developments in this area have explored the chromic responses of CLC elastomers to different types of mechanical activation. For example, the chromic transitions on applying uniaxial tension perpendicular to the helical axis were first examined experimentally by Cicuta, Tajbakhsh and Terentjev[132,133] and also later applied to mechanically tunable lasers in single[134] and layered[135,136] film structures. Most recently, Palffy-Muhoray and coworkers used compression along the helical axis, obtaining a system in which both pitch and lasing wavelength are linearly dependent on the strain applied to the film.[137] Parallel to this, it has also been found possible to induce colour changes from CLCs with force by applying a coating of the pure CLC on top of a polymeric material.[138]

Photonic polymer materials and cholesteric liquid crystals are arguably the most mature optical stress-sensing technologies accessible today as they have already been implemented in working devices. However, the optical response is achieved at the cost of significantly changing the material's overall properties and they therefore are maybe not the most universal probes available.

3.7 Conclusion

The study and development of polymers that change their optical properties under the influence of mechanical force is a thriving area of research that has resulted in the introduction of a range of different concepts that provide a change in absorption or emission intensity and wavelength. Some of these concepts are well underway to be developed into commercial technologies, such as the use of photonic band gap elastomers (Elast-Ink) as pressure sensitive materials for fingerprint detection or as a counterfeit measure in banknotes. More recently established methods, such as the use of mechanoluminescence, are at a much earlier stage of development, but promise unrivalled detectability and spatiotemporal resolution for the study of molecular damage mechanisms in polymeric materials. Complementary to these methods, the high force sensitivity of methods based on mechanochromism of conjugated polymers opens opportunities for the detection of weak forces in a biological context. Regardless of their stage of development, each operation mechanism has its preferred field of application, based on the extent to which the method is able to meet the specific requirement of the application. For instance, in monitoring long-term deformation and damage in end-user products, manufacturing cost, the long-term thermal stability and the photostability of the probe molecules are of paramount importance, while in fundamental studies of molecular scale forces, sensitivity is of prime concern.

Although the optical response of polymers to mechanical input is fundamentally a physical phenomenon, advances in the area are dependent on the development of synthetic molecules that embody the required photophysical effect, and their incorporation in polymers. Therefore, it should not come as a surprise that many of the key developments in the area originate in labs that bring together expertise in synthetic organic and polymer chemistry. For continued development of this exciting research area, a truly multidisciplinary approach is essential, taking input from materials science, optics and synthetic chemistry. Only with their combined expertise and imagination, scientists in these areas will be able to fulfil the potential of this fascinating class of materials.

Acknowledgements

The work was supported by the Ministry of Education, Culture and Science (Gravity program 024.001.035). J. M. C. was supported by the Council for Chemical Sciences of the Netherlands Organization for Scientific Research (CW-NWO), and R. G. was supported by the Deutsche Forschungsgemeinschaft (DFG) through a research fellowship – GO 2634/1-1.

References

1. C. J. Hellier, *Handbook of Nondestructive Evaluation*, McGraw-Hill Education, New York, 2nd edn, 2012.
2. M. M. Caruso, D. A. Davis, Q. Shen, S. A. Odom, N. R. Sottos, S. R. White and J. S. Moore, *Chem. Rev.*, 2009, **109**, 5755–5798.
3. D. R. T. Roberts and S. J. Holder, *J. Mater. Chem.*, 2011, **21**, 8256–8268.
4. Y. Jiang, *Mater. Sci. Eng., C*, 2014, **45**, 682–689.
5. Z. V. Todres, *J. Chem. Res.*, 2004, **2004**, 89–93.
6. G. Kaupp, *CrystEngComm*, 2009, **11**, 388–403.
7. Y. Sagara and T. Kato, *Nat. Chem.*, 2009, **1**, 605–610.
8. Z. Chi, X. Zhang, B. Xu, X. Zhou, C. Ma, Y. Zhang, S. Liu and J. Xu, *Chem. Soc. Rev.*, 2012, **41**, 3878–3896.
9. M. J. Jacobs and K. Blank, *Chem. Sci.*, 2014, **5**, 1680–1697.
10. Z. Ma, Z. Wang, M. Teng, Z. Xu and X. Jia, *ChemPhysChem*, 2015, **16**, 1811–1828.
11. Y. Sagara, S. Yamane, M. Mitani, C. Weder and T. Kato, *Adv. Mater.*, 2016, **28**, 1073–1095.

12. M. Montalti, A. Credi, L. Prodi and M. T. Gandolfi, *Handbook of Photochemistry*, CRC Press, 3rd edn, 2006.
13. *Photonic Materials for Sensing, Biosensing and Display Devices*, ed. M. J. Serpe, Y. Kang and Q. M. Zhang, Springer, Cham, vol. 229, 2016.
14. J. R. Lakowicz, *Principles of Fluorescence Spectroscopy*, Springer, New York, 3rd edn, 2006.
15. H. S. Rye, J. M. Dabora, M. A. Quesada, R. A. Mathies and A. N. Glazer, *Anal. Biochem.*, 1993, **208**, 144–150.
16. S. W. Hell and J. Wichmann, *Opt. Lett.*, 1994, **19**, 780.
17. S. W. Hell, *Science*, 2007, **316**, 1153–1158.
18. H. Dürr and H. Bouas-Laurent, *Photochromism: Molecules and Systems*, Elsevier Ltd, Oxford, 2nd edn, 2003.
19. S.-J. Kim and D. H. Reneker, *Polym. Bull.*, 1993, **31**, 367–374.
20. S. K. Surampudi, H. R. Patel, G. Nagarjuna and D. Venkataraman, *Chem. Commun.*, 2013, **49**, 7519–7521.
21. P. K. Kundu and R. Klajn, *ACS Nano*, 2014, **8**, 11913–11916.
22. K. L. Berkowski, S. L. Potisek, C. R. Hickenboth and J. S. Moore, *Macromolecules*, 2005, **38**, 8975–8978.
23. S. L. Potisek, D. A. Davis, N. R. Sottos, S. R. White and J. S. Moore, *J. Am. Chem. Soc.*, 2007, **129**, 13808–13809.
24. D. A. Davis, A. Hamilton, J. Yang, L. D. Cremar, D. Van Gough, S. L. Potisek, M. T. Ong, P. V. Braun, T. J. Martínez, S. R. White, J. S. Moore and N. R. Sottos, *Nature*, 2009, **459**, 68–72.
25. J. Li, C. Nagamani and J. S. Moore, *Acc. Chem. Res.*, 2015, **48**, 2181–2190.
26. B. A. Beiermann, D. A. Davis, S. L. B. Kramer, J. S. Moore, N. R. Sottos and S. R. White, *J. Mater. Chem.*, 2011, **21**, 8443–8447.
27. C. M. Kingsbury, P. A. May, D. A. Davis, S. R. White, J. S. Moore and N. R. Sottos, *J. Mater. Chem.*, 2011, **21**, 8381–8388.
28. G. O'Bryan, B. M. Wong and J. R. McElhanon, *ACS Appl. Mater. Interfaces*, 2010, **2**, 1594–1600.
29. G. I. Peterson, M. B. Larsen, M. A. Ganter, D. W. Storti and A. J. Boydston, *ACS Appl. Mater. Interfaces*, 2014, **7**, 577–583.
30. G. R. Gossweiler, G. B. Hewage, G. Soriano, Q. Wang, G. W. Welshofer, X. Zhao and S. L. Craig, *ACS Macro Lett.*, 2014, **3**, 216–219.
31. L.-J. Wang, X.-J. Zhou, X.-H. Zhang and B.-Y. Du, *Macromolecules*, 2016, **49**, 98–104.
32. X. Fang, H. Zhang, Y. Chen, Y. Lin, Y. Xu and W. Weng, *Macromolecules*, 2013, **46**, 6566–6574.
33. G. Hong, H. Zhang, Y. Lin, Y. Chen, Y. Xu, W. Weng and H. Xia, *Macromolecules*, 2013, **46**, 8649–8656.
34. Y. Chen, H. Zhang, X. Fang, Y. Lin, Y. Xu and W. Weng, *ACS Macro Lett.*, 2014, **3**, 141–145.
35. C. M. Degen, P. A. May, J. S. Moore, S. R. White and N. R. Sottos, *Macromolecules*, 2013, **46**, 8917–8921.
36. S. Jiang, L. Zhang, T. Xie, Y. Lin, H. Zhang, Y. Xu, W. Weng and L. Dai, *ACS Macro Lett.*, 2013, **2**, 705–709.

37. J. R. Hemmer, P. D. Smith, M. van Horn, S. Alnemrat, B. P. Mason, J. R. de Alaniz, S. Osswald and J. P. Hooper, *J. Polym. Sci., Part B: Polym. Phys.*, 2014, **52**, 1347–1356.
38. C. K. Lee, B. A. Beiermann, M. N. Silberstein, J. Wang, J. S. Moore, N. R. Sottos and P. V. Braun, *Macromolecules*, 2013, **46**, 3746–3752.
39. J. W. Kim, Y. Jung, G. W. Coates and M. N. Silberstein, *Macromolecules*, 2015, **48**, 1335–1342.
40. A.-D. N. Celestine, B. A. Beiermann, P. A. May, J. S. Moore, N. R. Sottos and S. R. White, *Polymer*, 2014, **55**, 4164–4171.
41. Q. Wang, G. R. Gossweiler, S. L. Craig and X. Zhao, *J. Mech. Phys. Solids*, 2015, **82**, 320–344.
42. G. R. Gossweiler, T. B. Kouznetsova and S. L. Craig, *J. Am. Chem. Soc.*, 2015, **137**, 6148–6151.
43. M.-J. Teng, X.-R. Jia, X.-F. Chen and Y. Wei, *Angew. Chem., Int. Ed.*, 2012, **51**, 6398–6401.
44. Z. Ma, M. Teng, Z. Wang, S. Yang and X. Jia, *Angew. Chem., Int. Ed.*, 2013, **52**, 12268–12272.
45. Z. Ma, Z. Wang, Z. Xu, X. Jia and Y. Wei, *J. Mater. Chem. C*, 2015, **3**, 3399–3405.
46. Z. Wang, Z. Ma, Y. Wang, Z. Xu, Y. Luo, Y. Wei and X. Jia, *Adv. Mater.*, 2015, **27**, 6469–6474.
47. H. Staudinger and W. Heuer, *Ber. Dtsch. Chem. Ges. A/B*, 1934, **67**, 1159–1164.
48. J. M. Clough, A. Balan and R. P. Sijbesma, in *Polymer Mechanochemistry*, ed. R. Boulatov, Springer International Publishing, 2015, pp. 209–238.
49. N. Yoshie, S. Saito and N. Oya, *Polymer*, 2011, **52**, 6074–6079.
50. S. S. M. Konda, J. N. Brantley, B. T. Varghese, K. M. Wiggins, C. W. Bielawski and D. E. Makarov, *J. Am. Chem. Soc.*, 2013, **135**, 12722–12729.
51. J. Li, T. Shiraki, B. Hu, R. A. E. Wright, B. Zhao and J. S. Moore, *J. Am. Chem. Soc.*, 2014, **136**, 15925–15928.
52. D. C. Church, G. I. Peterson and A. J. Boydston, *ACS Macro Lett.*, 2014, **3**, 648–651.
53. S. Billiet, K. De Bruycker, F. Driessen, H. Goossens, V. Van Speybroeck, J. M. Winne and F. E. Du Prez, *Nat. Chem.*, 2014, **6**, 815–821.
54. R. Göstl and R. P. Sijbesma, *Chem. Sci.*, 2016, **7**, 370–375.
55. Y.-K. Song, K.-H. Lee, W.-S. Hong, S.-Y. Cho, H.-C. Yu and C.-M. Chung, *J. Mater. Chem.*, 2011, **22**, 1380–1386.
56. Z. S. Kean, G. R. Gossweiler, T. B. Kouznetsova, G. B. Hewage and S. L. Craig, *Chem. Commun.*, 2015, **51**, 9157–9160.
57. C. R. Hickenboth, J. S. Moore, S. R. White, N. R. Sottos, J. Baudry and S. R. Wilson, *Nature*, 2007, **446**, 423–427.
58. J. Wang, T. B. Kouznetsova, Z. Niu, M. T. Ong, H. M. Klukovich, A. L. Rheingold, T. J. Martinez and S. L. Craig, *Nat. Chem.*, 2015, **7**, 323–327.
59. A. Löwenbein and H. Simonis, *Ber. Dtsch. Chem. Ges. A/B*, 1924, **57**, 2040–2048.

60. A. Löwenbein, *Ber. Dtsch. Chem. Ges. A/B*, 1925, **58**, 601–609.
61. A. Löwenbein and R. F. Gagarin, *Ber. Dtsch. Chem. Ges. A/B*, 1925, **58**, 2643–2644.
62. A. Löwenbein and H. Schmidt, *Ber. Dtsch. Chem. Ges. A/B*, 1927, **60**, 1851–1861.
63. K. Imato, A. Irie, T. Kosuge, T. Ohishi, M. Nishihara, A. Takahara and H. Otsuka, *Angew. Chem., Int. Ed.*, 2015, **54**, 6168–6172.
64. K. Imato, T. Kanehara, T. Ohishi, M. Nishihara, H. Yajima, M. Ito, A. Takahara and H. Otsuka, *ACS Macro Lett.*, 2015, **4**, 1307–1311.
65. Y. Chen, A. J. H. Spiering, S. Karthikeyan, G. W. M. Peters, E. W. Meijer and R. P. Sijbesma, *Nat. Chem.*, 2012, **4**, 559–562.
66. E. Ducrot, Y. Chen, M. Bulters, R. P. Sijbesma and C. Creton, *Science*, 2014, **344**, 186–189.
67. Z. S. Kean, J. L. Hawk, S. Lin, X. Zhao, R. P. Sijbesma and S. L. Craig, *Adv. Mater.*, 2014, **26**, 6013–6018.
68. Y. Chen and R. P. Sijbesma, *Macromolecules*, 2014, **47**, 3797–3805.
69. J. M. Clough and R. P. Sijbesma, *ChemPhysChem*, 2014, **15**, 3565–3571.
70. G. B. Schuster, N. J. Turro, H. C. Steinmetzer, A. P. Schaap, G. Faler, W. Adam and J. C. Liu, *J. Am. Chem. Soc.*, 1975, **97**, 7110–7118.
71. J. M. Clough, A. Balan, T. L. J. van Daal and R. P. Sijbesma, *Angew. Chem., Int. Ed.*, 2016, **55**, 1445–1449.
72. R. W. Carpick, D. Y. Sasaki, M. S. Marcus, M. A. Eriksson and A. R. Burns, *J. Phys.: Condens. Matter*, 2004, **16**, R679.
73. M. A. Reppy and B. A. Pindzola, *Chem. Commun.*, 2007, 4317–4338.
74. O. Yarimaga, J. Jaworski, B. Yoon and J.-M. Kim, *Chem. Commun.*, 2012, **48**, 2469–2485.
75. R. A. Nallicheri and M. F. Rubner, *Macromolecules*, 1991, **24**, 517–525.
76. R. A. Koevoets, S. Karthikeyan, P. C. M. M. Magusin, E. W. Meijer and R. P. Sijbesma, *Macromolecules*, 2009, **42**, 2609–2617.
77. H. Feng, J. Lu, J. Li, F. Tsow, E. Forzani and N. Tao, *Adv. Mater.*, 2013, **25**, 1729–1733.
78. D.-H. Park, J. Hong, I. S. Park, C. W. Lee and J.-M. Kim, *Adv. Funct. Mater.*, 2014, **24**, 5186–5193.
79. Y. Tomioka, N. Tanaka and S. Imazeki, *J. Chem. Phys.*, 1989, **91**, 5694–5700.
80. R. W. Carpick, D. Y. Sasaki and A. R. Burns, *Langmuir*, 2000, **16**, 1270–1278.
81. A. R. Burns, R. W. Carpick, D. Y. Sasaki, J. A. Shelnutt and R. Haddad, *Tribol. Lett.*, 2001, **10**, 89–96.
82. K. Becker and J. M. Lupton, *J. Am. Chem. Soc.*, 2005, **127**, 7306–7307.
83. E. Da Como, K. Becker, J. Feldmann and J. M. Lupton, *Nano Lett.*, 2007, **7**, 2993–2998.
84. H. E. Cingil, I. M. Storm, Y. Yorulmaz, D. W. te Brake, R. de Vries, M. A. Cohen Stuart and J. Sprakel, *J. Am. Chem. Soc.*, 2015, **137**, 9800–9803.
85. J. Kim and T. M. Swager, *Nature*, 2001, **411**, 1030–1034.

86. A. Pucci and G. Ruggeri, *J. Mater. Chem.*, 2011, **21**, 8282–8291.
87. F. Ciardelli, G. Ruggeri and A. Pucci, *Chem. Soc. Rev.*, 2013, **42**, 857–870.
88. C. Löwe and C. Weder, *Adv. Mater.*, 2002, **14**, 1625–1629.
89. B. R. Crenshaw and C. Weder, *Chem. Mater.*, 2003, **15**, 4717–4724.
90. J. Kunzelman, B. R. Crenshaw, M. Kinami and C. Weder, *Macromol. Rapid Commun.*, 2006, **27**, 1981–1987.
91. B. R. Crenshaw, M. Burnworth, D. Khariwala, A. Hiltner, P. T. Mather, R. Simha and C. Weder, *Macromolecules*, 2007, **40**, 2400–2408.
92. J. Lott and C. Weder, *Macromol. Chem. Phys.*, 2010, **211**, 28–34.
93. J. Kunzelman, M. Gupta, B. R. Crenshaw, D. A. Schiraldi and C. Weder, *Macromol. Mater. Eng.*, 2009, **294**, 244–249.
94. M. Kinami, B. R. Crenshaw and C. Weder, *Chem. Mater.*, 2006, **18**, 946–955.
95. B. R. Crenshaw and C. Weder, *Macromolecules*, 2006, **39**, 9581–9589.
96. A. Pucci, M. Bertoldo and S. Bronco, *Macromol. Rapid Commun.*, 2005, **26**, 1043–1048.
97. A. Pucci, C. Cappelli, S. Bronco and G. Ruggeri, *J. Phys. Chem. B*, 2006, **110**, 3127–3134.
98. A. Pucci, F. D. Cuia, F. Signori and G. Ruggeri, *J. Mater. Chem.*, 2007, **17**, 783–790.
99. S. Bao, J. Li, K. I. Lee, S. Shao, J. Hao, B. Fei and J. H. Xin, *ACS Appl. Mater. Interfaces*, 2013, **5**, 4625–4631.
100. F. Cellini, S. Khapli, S. D. Peterson and M. Porfiri, *Appl. Phys. Lett.*, 2014, **105**, 061907.
101. F. Cellini, L. Zhou, S. Khapli, S. D. Peterson and M. Porfiri, *Mech. Mater.*, 2016, **93**, 145–162.
102. A. Pucci, F. Donati, G. Ruggeri and F. Ciardelli, *e-Polym.*, 2013, **9**, 716–726.
103. F. Donati, A. Pucci, C. Cappelli, B. Mennucci and G. Ruggeri, *J. Phys. Chem. B*, 2008, **112**, 3668–3679.
104. F. Donati, A. Pucci and G. Ruggeri, *Phys. Chem. Chem. Phys.*, 2009, **11**, 6276–6282.
105. S. Karthikeyan and R. P. Sijbesma, *Macromolecules*, 2009, **42**, 5175–5178.
106. A.-Y. Jee and M. Lee, *Chem. Phys. Lett.*, 2011, **501**, 287–291.
107. J. R. Kumpfer, S. D. Taylor, W. B. Connick and S. J. Rowan, *J. Mater. Chem.*, 2012, **22**, 14196–14204.
108. A. Lavrenova, J. Farkas, C. Weder and Y. C. Simon, *ACS Appl. Mater. Interfaces*, 2015, **7**, 21828–21834.
109. J. Mei, N. L. C. Leung, R. T. K. Kwok, J. W. Y. Lam and B. Z. Tang, *Chem. Rev.*, 2015, **115**, 11718–11940.
110. G. Iasilli, A. Battisti, F. Tantussi, F. Fuso, M. Allegrini, G. Ruggeri and A. Pucci, *Macromol. Chem. Phys.*, 2014, **215**, 499–506.
111. S.-J. Yoon, J. W. Chung, J. Gierschner, K. S. Kim, M.-G. Choi, D. Kim and S. Y. Park, *J. Am. Chem. Soc.*, 2010, **132**, 13675–13683.
112. Y. Wu, J. Hu, H. Huang, J. Li, Y. Zhu, B. Tang, J. Han and L. Li, *J. Polym. Sci., Part B: Polym. Phys.*, 2014, **52**, 104–110.

113. J. Ge and Y. Yin, *Angew. Chem., Int. Ed.*, 2011, **50**, 1492–1522.
114. E. P. Chan, J. J. Walish, A. M. Urbas and E. L. Thomas, *Adv. Mater.*, 2013, **25**, 3934–3947.
115. C. Fenzl, T. Hirsch and O. S. Wolfbeis, *Angew. Chem., Int. Ed.*, 2014, **53**, 3318–3335.
116. H. Li, X. Sun and H. Peng, *ChemPhysChem*, 2015, **16**, 3761–3768.
117. S. A. Asher, J. Holtz, L. Liu and Z. Wu, *J. Am. Chem. Soc.*, 1994, **116**, 4997–4998.
118. J. M. Jethmalani and W. T. Ford, *Chem. Mater.*, 1996, **8**, 2138–2146.
119. J. M. Jethmalani, H. B. Sunkara, W. T. Ford, S. L. Willoughby and B. J. Ackerson, *Langmuir*, 1997, **13**, 2633–2639.
120. Y. F. Yue, M. A. Haque, T. Kurokawa, T. Nakajima and J. P. Gong, *Adv. Mater.*, 2013, **25**, 3106–3110.
121. Y. Yue, T. Kurokawa, M. A. Haque, T. Nakajima, T. Nonoyama, X. Li, I. Kajiwara and J. P. Gong, *Nat. Commun.*, 2014, **5**, 4659.
122. X.-Q. Wang, C.-F. Wang, Z.-F. Zhou and S. Chen, *Adv. Opt. Mater.*, 2014, **2**, 652–662.
123. Y. Cho, S. Y. Lee, L. Ellerthorpe, G. Feng, G. Lin, G. Wu, J. Yin and S. Yang, *Adv. Funct. Mater.*, 2015, **25**, 6041–6049.
124. X. Jia, J. Wang, K. Wang and J. Zhu, *Langmuir*, 2015, **31**, 8732–8737.
125. X. Sun, J. Zhang, X. Lu, X. Fang and H. Peng, *Angew. Chem., Int. Ed.*, 2015, **54**, 3630–3634.
126. D. Yang, S. Ye and J. Ge, *Chem. Commun.*, 2015, **51**, 16972–16975.
127. M. O'Neill and S. M. Kelly, *Adv. Mater.*, 2003, **15**, 1135–1146.
128. H. Coles and S. Morris, *Nat. Photonics*, 2010, **4**, 676–685.
129. H. Finkelmann, S. T. Kim, A. Muñoz, P. Palffy-Muhoray and B. Taheri, *Adv. Mater.*, 2001, **13**, 1069–1072.
130. J. Schmidtke, S. Kniesel and H. Finkelmann, *Macromolecules*, 2005, **38**, 1357–1363.
131. S. T. Kim and H. Finkelmann, *Macromol. Rapid Commun.*, 2001, **22**, 429–433.
132. P. Cicuta, A. R. Tajbakhsh and E. M. Terentjev, *Phys. Rev. E*, 2002, **65**, 051704.
133. P. Cicuta, A. R. Tajbakhsh and E. M. Terentjev, *Phys. Rev. E*, 2004, **70**, 011703.
134. P. V. Shibaev, P. Rivera, D. Teter, S. Marsico, M. Sanzari, V. Ramakrishnan and E. Hanelt, *Opt. Express*, 2008, **16**, 2965.
135. Y. Hirota, Y. Ji, F. Serra, A. R. Tajbakhsh and E. M. Terentjev, *Opt. Express*, 2008, **16**, 5320.
136. F. Serra, M. A. Matranga, Y. Ji and E. M. Terentjev, *Opt. Express*, 2010, **18**, 575.
137. A. Varanytsia, H. Nagai, K. Urayama and P. Palffy-Muhoray, *Sci. Rep.*, 2015, **5**, 17739.
138. O. T. Picot, M. Dai, E. Billoti, D. J. Broer, T. Peijs and C. W. M. Bastiaansen, *RSC Adv.*, 2013, **3**, 18794–18798.

CHAPTER 4

Materials Design Principles for Mechanochemical Transduction

M. N. SILBERSTEIN* AND M. S. MANIVANNAN

Cornell University, Sibley School of Mechanical and Aerospace Engineering, Ithaca, NY, 14850, USA
*E-mail: ms2682@cornell.edu

4.1 Introduction

Productive mechanochemistry can be designed in polymers and composites through the covalent incorporation of mechanophores – chemical units that undergo a specific chemical transformation in response to applied force.[1] Force is transmitted to the molecular (mechanophore) scale through the host material and, therefore, the process of force transmission from the macroscale to the molecular scale is highly polymer dependent. In addition to transmitting the force, the host matrix can also inhibit the response. For example, intuitively one would expect a conformation change to occur more slowly in a surrounding body of higher rather than lower viscosity. It is useful to categorize the mechanochemically responsive (MCR) materials into elastomers, glassy polymers and composites, given the distinct mechanisms each has for accommodating deformation.

In this chapter, we will first present preparatory information: mechanics terminology that is needed to examine mechanochemical transduction in solid-state polymers, an overview of mechanophore kinetics including the basic equations and typical experimental methods along with their relative merits.

Polymer Chemistry Series No. 26
Mechanochemistry in Materials
Edited by Yoan C. Simon and Stephen L. Craig

Published by the Royal Society of Chemistry, www.rsc.org

We then address the existing experimental results and theory for mechanophore triggering within bulk polymers in terms of material class: elastomers, glassy and composites. Finally, we discuss the current state-of-the-art in mechanochemically adaptive polymer networks and prospects for advancing the field.

4.2 Mechanics Terminology

Solid mechanics is often essential for understanding and designing MCR polymers. Here, we review a few key terms and concepts that will be used throughout the chapter.

In a one-dimensional mechanics perspective, both stress and deformation can be described by scalar values. Strain and stretch are both measures of deformation. Stretch, most commonly used for materials that can deform to many times their initial length, is defined as:

$$\lambda = \frac{l}{l_0} \tag{4.1}$$

where l_0 and l are the initial and instantaneous lengths of the specimen, respectively. Engineering strain is defined as:

$$\varepsilon_{\text{eng}} = \frac{l - l_0}{l_0} = \lambda - 1 \tag{4.2}$$

For large deformations, we define a true strain that is the integral of strain increments:

$$\varepsilon_{\text{true}} = \int \frac{\partial l}{l_0} = \ln(\lambda) \tag{4.3}$$

Strain rate is the rate at which strain is applied and can be given in either engineering or true strain. True strain rate is the more physically relevant value but, in practice, engineering strain rate is used for simplicity of implementation. All strains and strain rates referred to in this chapter are engineering unless explicitly denoted as true.

Stress is force normalized by the area over which the force is acting. The sign convention for stress is that positive means tensile, although often the signs are interchanged when only one direction of loading is applied within a single experiment. In large deformation, we distinguish between engineering stress (σ_{eng}) and true stress (σ_{true}) where engineering stress is divided by original area and true stress is divided by the instantaneous area. For uniaxial or compressive loading of an incompressible material, engineering and true stress are related by:

$$\sigma_{\text{true}} = \sigma_{\text{eng}}(1 + \varepsilon_{\text{eng}}) \tag{4.4}$$

Stresses referred to in this chapter are engineering unless explicitly denoted as true.

In three-dimensional space, deformation is described in terms of the deformation gradient (**F**).

$$\mathbf{F} = \frac{\partial \mathbf{x}}{\partial \mathbf{X}} \tag{4.5}$$

where **X** and **x** are the coordinates in the initial and deformed bodies respectively. The deformation gradient is a second-order tensor that describes volume change, shape change and rotation of the body. The stretching part is captured by the product:

$$\mathbf{U} = \sqrt{\mathbf{C}} = \sqrt{\mathbf{F}^{\mathrm{T}}\mathbf{F}} \tag{4.6}$$

where **U** is the right stretch tensor and **C** is the right Cauchy–Green tensor and also by:

$$\mathbf{V} = \sqrt{\mathbf{B}} = \sqrt{\mathbf{F}\mathbf{F}^{\mathrm{T}}} \tag{4.7}$$

where **V** is the left stretch tensor and **B** is the left Cauchy–Green tensor. The eigenvalues of **U** and **V** are the principal stretches λ_1, λ_2, λ_3. The ratio of the current volume to the initial volume is captured by:

$$J = \det \mathbf{F} = \lambda_1\lambda_2\lambda_3 \tag{4.8}$$

For instance, for uniaxial tension of an incompressible material, λ_1 is the applied stretch and $\lambda_2 = \lambda_3 = \lambda_1{}^{1/2}$. The deformation of polymers is also frequently described in terms of the first invariant:

$$I_1 = \mathrm{trace}(\mathbf{B}) = \lambda_1{}^2 + \lambda_2{}^2 + \lambda_3{}^2 \tag{4.9}$$

Physically, the first invariant is a good approximation of the effective stretch (magnitude, not direction) felt by the material.

Stress (**T**) is also a tensor. Physically it is useful to think of stress in terms of a hydrostatic portion and a deviatoric (shear) portion. The hydrostatic portion of stress is the effective pressure – the component of stress that is the same magnitude in every normal direction.

$$p = -\sigma_{\mathrm{mean}} = \frac{1}{3}\mathrm{trace}(\mathbf{T}) \tag{4.10}$$

The remaining portion of the stress is the deviatoric stress (**T′**):

$$\mathbf{T}' = \mathbf{T} - p\mathbf{1} \tag{4.11}$$

where **1** is the second-order identity matrix.

Constitutive models are often prescribed from a strain energy density function (ψ) to ensure compatibility with the laws of thermodynamics. The stress is taken as the derivative of the strain energy function by:

$$\mathbf{T} = \frac{1}{J}\frac{\partial \bar{\psi}}{\partial \mathbf{F}}\mathbf{F}^{\mathrm{T}} = \frac{2}{J}\left[\left(\frac{\partial \hat{\psi}}{\partial I_1} + I_1\frac{\partial \hat{\psi}}{\partial I_2}\right)\mathbf{B} - \frac{\partial \hat{\psi}}{\partial I_2}\mathbf{B}^2\right] + \frac{\partial \hat{\psi}}{\partial J}\mathbf{1} \tag{4.12}$$

where I_2 is the second invariant of $\mathbf{B}$, *bar* indicates that the strain energy is expressed as a function of the deformation gradient and *hat* indicates that the strain energy is expressed as a function of stretch invariants.

4.3 Mechanophore Kinetics

For the purposes of this chapter, we will accept the Bell theory picture of the mechanophore reaction and discuss mechanophore responsiveness in the context of transition state theory.[2,3] The mechanophore can be viewed as a molecular unit with a vector defining its orientation. When force is applied across the mechanophore vector, the force does work on the mechanophore and reduces the energy barrier (ΔG_{f}) for transitioning from the closed/inactive state to the open/active state (see Figure 4.1). The terms "closed" and "open" are often used in the context of electrocyclic reactions, and many mechanophores are based on such reactions, but we adopt a convention in this chapter in which "closed" and "open" can be thought of more generally to represent the inactive and activated forms of a given mechanophore.

$$\Delta G_{\mathrm{f}} = \Delta G_{\mathrm{f}}^{0} - f\Delta x_{\mathrm{f}} \tag{4.13}$$

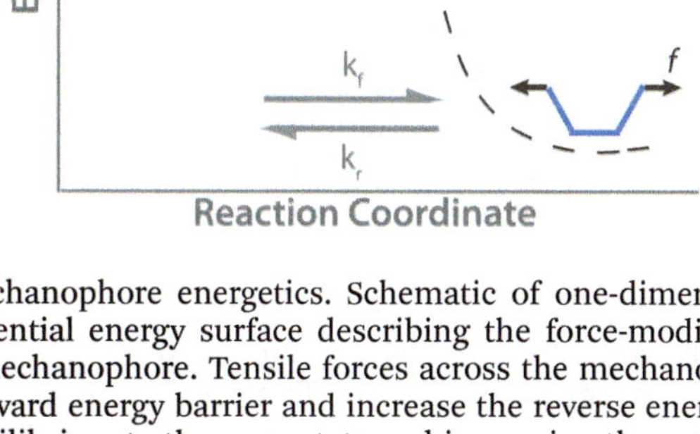

Figure 4.1 Mechanophore energetics. Schematic of one-dimensional version of potential energy surface describing the force-modified energetics of a mechanophore. Tensile forces across the mechanophore reduce the forward energy barrier and increase the reverse energy barrier driving equilibrium to the open state and increasing the rate of reaction from closed to open.

where ΔG_f^0 is the force-free energy barrier, f is the magnitude of the applied force vector and x_f is the reaction coordinate along the mechanophore orientation vector in the closed state ($\Delta x_f > 0$, other factors such as light could be added to the energy barrier function as well). This force-dependent energy barrier determines the rate of activation (k_f, transition from the closed to the open state) at a given force and temperature.[4]

$$k_f = \frac{k_b \theta}{h} \frac{Q_f^+}{Q} \exp\left[\frac{-\Delta G_f}{k_b \theta}\right] \tag{4.14}$$

where k_b is Boltzmann's constant, θ is the absolute temperature and Q_f^+/Q is the forward attempt frequency. If the mechanophore reaction is reversible, then there is also a force-modified reverse energy barrier (ΔG_r) and reaction rate (k_r).

$$\Delta G_r = \Delta G_r^0 - f \Delta x_r \tag{4.15}$$

$$k_r = \frac{k_b \theta}{h} \frac{Q_r^+}{Q} \exp\left[\frac{-\Delta G_r}{k_b \theta}\right] \tag{4.16}$$

where ΔG_r^0 is the force-free energy barrier to the reverse reaction, x_r is the reaction coordinate along the mechanophore orientation vector in the open state ($\Delta x_r < 0$) and Q_r^+/Q is the reverse attempt frequency. The overall rate of activation of a population of mechanophores subjected to a uniform force is then given by:

$$\dot{\alpha} = k_f(1 - \alpha) + k_r \alpha \tag{4.17}$$

where α is the fraction of mechanophores in the open state and $\dot{\alpha}$ is the rate of activation. For an irreversible reaction, k_r is zero. Eqn (4.17) can also be used to find the equilibrium mechanophore state by setting $\dot{\alpha} = 0$. For an irreversible mechanophore, the equilibrium state is fully activated but, practically speaking, this state will not be reached in the absence of force on a laboratory time scale unless the mechanophore is thermally unstable enough to make it impractical.

There are two key aspects from this reaction rate perspective to take with us as we move to discussing mechanophores in MCR polymers. The first is that the rate constants are exponentially dependent on force. For any set of mechanophore energetics, there is a range of force over which the activation response changes drastically on the laboratory time scale (see Figure 4.2a). Though ill-quantified, force fluctuations are well established to exist in polymers. Below this force, the activation rate is too small to notice and above this force range the activation is effectively instantaneous. This exponential dependence also means that small fluctuations in force, while holding the mean force constant, may result in large increases in mechanophore

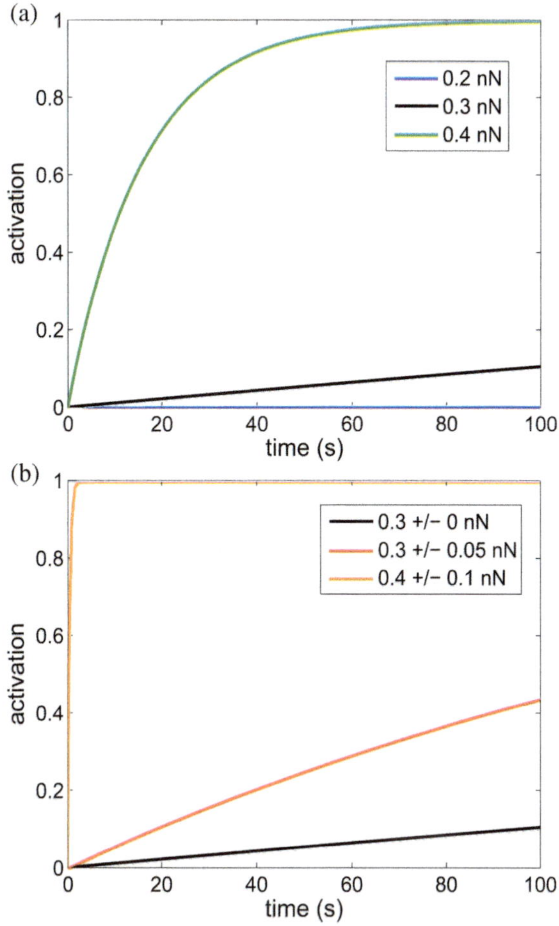

Figure 4.2 Activation *versus* time for a population of mechanophores subjected to uniform force. (a) Constant force. (b) Force fluctuating about a mean value, standard deviation indicated in legend.

activation (see Figure 4.2b). In polymers, force distributions are the norm. The second key aspect of mechanophore kinetics is that from a mechanics perspective, mechanophore reactivity happens along a vector. Only the dot product of the force and the reaction coordinate matter – this is a scalar quantity that modifies the energy barrier on the mechanophore scale. While force and/or displacement are also applied on a bulk polymer scale, the polymer experiences the externally applied force and displacement in terms of internal stresses and strains. We will discuss how externally applied deformation becomes mechanophore-scale force in subsequent sections; for now, it is sufficient to point out that neither force nor stress magnitude at the bulk scale can be directly substituted into eqn (4.13) and (4.15) to find the mechanophore rate of activation.

4.4 Experimental Techniques for MCR Polymers

Experiments are essential for advancing design of mechanochemically responsive polymers. Both experimental design and results analysis strongly influence the knowledge gained from characterizing a particular mechanophore-polymer system. Mechanophores with optically measurable responses are by far the most frequently used within bulk polymers because it is relatively straightforward to monitor mechanophore state during mechanical deformation. The two demonstrated options for optically measurable responses involve color change, and luminescence. Here we will briefly introduce the relevant mechanophores in the solid-state polymer literature (discussed in depth in other chapters), discuss in some detail experimental methodologies for characterizing bulk polymers with optically measurable mechanophores and briefly survey examples of non-optical detection of mechanophore activation.

Spiropyran (SP) is a mechanophore that has both an absorption and a fluorescence change associated with its transition to the open (merocyanine) state (see Figure 4.3). The ability to mechanically shift the equilibrium of spiropyran to merocyanine was discovered in the seminal work of the Sottos group.[5] This mechanophore is reversible and can also be manipulated by light and heat. UV light and heat drive the mechanophore population towards the merocyanine state. Visible light drives the mechanophore towards the SP state.[6] SP has a small absorption in the yellow part of the visible spectrum and no fluorescence response. Merocyanine can take on a red, purple or blue color depending on the host matrix and emits at 600 to 650 nm when excited with a 532 nm laser.[5–8] SP can be covalently incorporated into a polymer through multiple attachment point options.[9,10] If the combination of attachment points transmits force across to a C–O bond, then SP is considered a mechanophore, otherwise it is considered a difunctional control. If SP is only covalently linked at one point, then it is called a monofunctional control – these controls tend to be less representative than difunctional controls since they are inherently side groups rather than forming a link in the backbone or cross-link. The choice of attachment points does change the force required to trigger

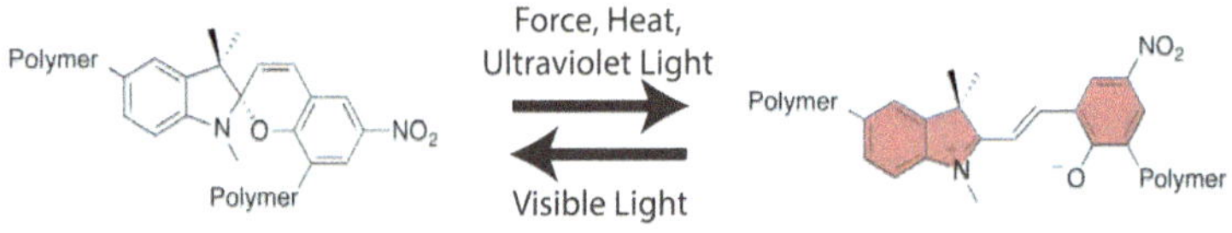

Figure 4.3 Spiropyran mechanophore. Ring opening of the spiropyran mechanophore can be induced by force applied at the indicated polymer linkage points, converting the mechanophore to the merocyanine form that has a strong fluorescent signature. This ring opening can also be induced by exposure to ultra violet light and reversed by exposure to visible light.

the mechanophore,[9] but we will not distinguish among them in this chapter, since SPs with different attachment points have not yet been compared within the same bulk polymer.

Dioxetanes are the most used mechanoluminescent mechanophore (see Figure 4.4). Bis(adamantly)-1,2-dioxetane as a mechanophore was first demonstrated by the Sijbesma group.[11] The mechanophore emits blue light upon activation. The emitted color can be modified with the addition of an acceptor molecule. The delay between mechanophore triggering and light emission is on the nanosecond scale. A key feature of mechanoluminescence is the transient nature of the light – light indicates the mechanophore transition from closed to open rather than the current mechanophore population state. This mechanophore transformation is scission based and therefore irreversible.

Several other mechanophores with associated optical changes have been implemented within solid-state polymers. Π-extended anthracene is a fluorescent mechanophore with a high intensity per active mechanophore (high quantum yield).[12] It is excited by and emits within the visible spectrum, making it experimentally convenient. Unlike SP, this mechanophore is scission based and irreversible. Similarly, a dimeric anthracene moiety emits in the 500–600 nm range when excited with UV light.[13] Diarylbibenzofuranone (DABBF) is a dynamic covalent mechanophore that transitions between a pale yellow when intact and blue when separated.[14] DABBF is distinct from the other mechanophores introduced thus far because its activation results in chain scission, but the mechanophore halves can then be reconnected either to each other or to other activated mechanophore halves.

The default characterization of MCR polymers is a uniaxial tensile test. In a uniaxial tensile test, either a dogbone or a high-aspect-ratio rectangular specimen is used to generate a state of uniaxial stress at the vertical center (gauge region) of the specimen. The specimen is then extended at a controlled grip displacement rate while the force is measured. It is good practice to optically measure the strain as well, particularly for elastomeric specimens at

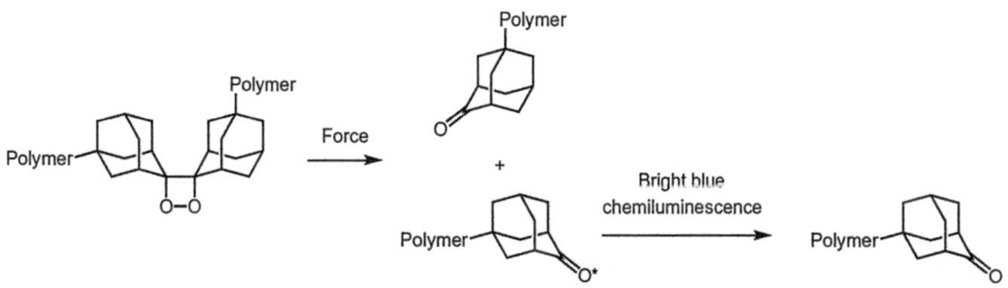

Figure 4.4 Dioxetane mechanophore. Bis(adamantyl)-1-2-dioxetane is a mechanophore that undergoes an irreversible scission when force is applied across the marked polymer attachment points and emits blue light upon relaxation from the excited state. Reprinted by permission from Macmillan Publishers Ltd: *Nature Chemistry* (ref. 11). Copyright 2012.

large strain for which out-of-gauge deformation is typically significant.[15] If the crosshead displacement is used to calculate strain, then the strain is most likely overestimated. Machine compliance can also obscure crosshead-based strain measurements, but tends to be a negligible factor in tensile testing unless operating near the load capacity of the machine. Tensile testing is compatible with *in-situ* optical characterization because there is a surface of the specimen exposed throughout the experiment (see Figure 4.5). For color change detection, all that is needed in addition to the standard tensile testing equipment is a color camera. The frame rate of the camera determines the time resolution of the experiment, but for quasi-static testing one to two Hertz is sufficient since the opto-mechanical state of the material changes slowly. A true white target next to the specimen increases the accuracy of the data by providing a means for correcting for lighting effects. Detection of mechanoluminescence is similar to that of color change, but the quality of the experiment improves considerably when the camera has higher light sensitivity and faster frame rate. Since the light intensity from each emission is transient, the signal will in theory change on the scale of a nanosecond. Longer exposure times simply integrate the luminescence at the cost of temporal resolution. Fluorescence measurements require additional optical components, but are typically more sensitive than color change measurements. First, an incident light is used to excite the fluorescence. Band-pass filters are then needed in front of the camera so that the camera only records intensity in the fluorescence band as the incident light will have greater intensity than the fluorescence. The camera for fluorescence detection need not resolve spectral intensities, since the wavelength filtering is done in front of the camera lens. With all three of these techniques, it is essential to keep the center of the specimen in the imaging window. The specimen center can be kept in the field of view by (1) moving the camera at half the displacement rate of the grip, (2) using a dual actuator load frame or (3) using a large enough field of view to see the specimen at its largest deformation. The third option is the simplest but causes a reduction of signal intensity and spatial resolution by reducing the number of the camera's pixels that are exposed to the regions

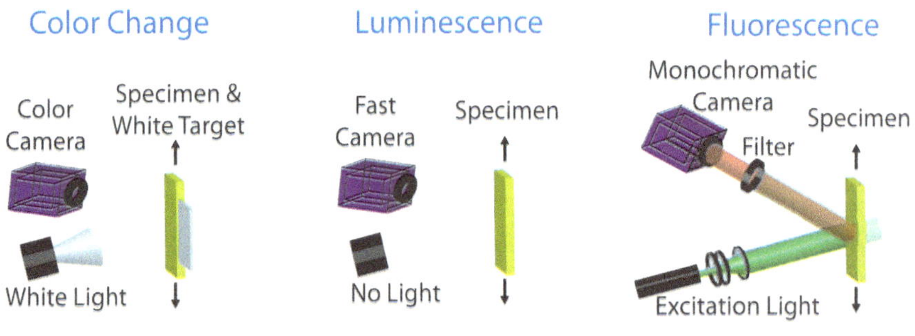

Figure 4.5 Experimental setups for tensile testing of optically responsive mechanophore-linked solid-state polymers.

of interest. In the absence of artifacts from sample deformation, both signal intensities (color and luminescence) are related to the number of mechanophores. If one were imaging a dilute solution containing mechanophores, for example, then the proportionality would be one to one; however, this proportionality is not guaranteed in the solid state. For instance, in a thick specimen, the color change on the back surface of the specimen will not be visible if the color change on the front surface of the specimen causes the specimen to be opaque. Under tensile deformation, the number of mechanophores per surface area decreases. Some authors therefore perform a thickness correction to adjust the intensity signal to the equivalent value for a non-changing number of mechanophores.[16,17] The validity of this correction is unclear at this time and likely depends on the initial sample dimensions, strain and degree of activation.

Fluorescence and color change signals can also be collected as spectrum of intensity *versus* wavelength rather than a single integrated intensity value. UV-Vis spectroscopy can be performed in either absorption or reflection mode to detect color. Rather than a single intensity value, spectroscopy provides intensity as a function of wavelength. Similarly, spectrofluorimetry can be performed to acquire fluorescence intensity as a function of wavelength for a given excitation wavelength. Obtaining an excitation and emission spectrum of a given mechanophore helps inform which band-pass filter should be chosen to maximize signal intensity and signal to noise in applications of fluorescence analysis. However, since UV-Vis experiments provide single-point spectra, spatial resolution is lost unless the specimen is rastered. It is therefore much more common to use a camera during the mechanical test and verify the full spectrum before and after the mechanical test.

Torsion, compression and plane strain tension are all less commonly used, but still quite useful experiments for MCR polymers. Torsion applies primarily shear deformation to the material. Torsion tests have the advantage of maintaining a constant gauge length and area, eliminating the need for a thickness correction and keeping the gauge in the field of view. However, hollow-cylinder torsion is difficult in terms of specimen preparation and solid-cylinder torsion is complex in terms of extracting both the stress and activation response. Because the shear strain is linearly proportional to the radial distance from the center, the specimen goes through yield and activates progressively from the outer surface to the center in a manner that is not readily invertible. The torsion specific mechanics at least can be mimicked with a finite element simulation to extract the polymer constitutive behavior. Compression testing is similar to tensile testing except that the specimen length to width ratio is typically much smaller in order to avoid bending. From a mechanics perspective, compression loading has a positive hydrostatic pressure component that is not present in tensile loading. From an imaging perspective, there can be difficulty with the compression plates shadowing the specimen and the specimen thickness tends to be too large for picking up fluorescence or color change through the depth. Plane-strain tension is performed identically to a tension test except the gauge region is much wider than it is long. This induces a biaxial stress state in the plane of the specimen that can be used to probe stress dependencies of the mechanochemical response. Plane-strain tension can be better than uniaxial tension for thin film specimens because the extra dimensional constraint inhibits specimen wrinkling.

Either the applied grip displacement or (less commonly) force history can be controlled throughout each of these loading modes. Monotonic deformation to failure at a constant strain rate is a good starting point for characterizing the mechanochemical response of a material. Additional information can then be gained through stress relaxation, creep and cyclic loading. For stress relaxation testing the material is quickly stretched to a set value of strain and then held at constant strain for an extended period of time while force and mechanophore activation are monitored. For creep testing, the material is quickly subjected to a set value of stress and then held at that stress for an extended period of time while strain and activation are monitored. While creep is defined as constant stress, in practice, constant force or torque is applied. Depending on the specimen geometry and constitutive behavior of the material, the stress may actually be changing with time even though the applied load does not change. When comparing simulations and experiments, it is therefore important to simulate the actual loading conditions. Cyclic loading consists of loading, unloading and reloading at least once and is typically performed under displacement control. There are no limits or guides on the number of cycles, stretch of each cycle and whether the maximum stretch should increase from cycle to cycle. Naturally, cycles well below the strain at which mechanophore activation initiates are of minimal value. Cyclic loading data are highly informative with respect to the mechanical properties of the material and provides a difficult (and useful) validation of any theoretical understanding of activation.

Another common means of conducting experiments on MCR polymers is to use a pellet press. The pellet press test is similar to a compression test; however, the experiment is not conducted at a controlled strain rate (important since both polymers and chemical reactions are time-dependent), strain is unknown, the pressure readout assumes that the specimen occupies the full area of the compression plates and the lip around the bottom plate can induce hydrostatic pressure or, alternatively, shear if there is a gap between the top plate and the lip. While there is a lot of valuable data published with pellet presses in the mechanochemistry field, and we include those results in our discussions below, these results do need to be interpreted with care, especially with respect to quantitative conclusions. For instance, when a polymer is folded and pressed again to the same pressure value the polymer is in fact being subjected to higher stress since it has a lower area.

MCR polymers with non-optical responses typically require *ex-situ* analysis to determine percent mechanophore activation, although *in-situ* chemical probing could also be successful in some cases. The appropriate experimental approach depends on both the mechanophore and the polymer matrix. For instance, Raman spectroscopy can be used to identify changes in the mechanophore state, but only if the polymer signal does not have peaks at the same location. Methods that will be mentioned here include NMR spectroscopy, electron paramagnetic resonance (EPR), Raman spectroscopy, Fourier-transform infrared spectroscopy (FTIR), size-exclusion chromatography (SEC), differential scanning calorimetry (DSC) and gas chromatography-mass spectrometry (GC-MS). NMR, EPR, GC-MS, Raman and FTIR all produce spectra that can be used to identify the changes in the chemical structure within the polymer. Microscope-compatible tensile stages should enable *in-situ* characterization of some non-optical mechanophore-linked polymers *via* Raman and FTIR moving forward. SEC identifies the molecular weight of polymers. DSC can be used to identify mechanophore activation for thermally triggerable mechanophores with strong endo- or exotherms. Our discussion in the remaining sections will focus on results rather than analysis methodology.

4.5 Elastomers

Elastomers are rubber-like synthetic materials made of a network of high-molecular-weight polymer chains. Under uniaxial tensile loading, the nominal stress response of elastomers follows a characteristic S-shaped curve – initial stiffness, relatively flat region and then strain hardening to failure.[18] The initial stiffness is in the 0.1–10 MPa range, and the failure strain is typically from 100% to 1000%.[18] Elastomers by definition exist above their glass transition temperature. While the ideal elastomer has no rate dependence, no hysteresis under cyclic loading and no anelastic deformation, many systems commonly referred to as elastomers do exhibit these traits.

The building block of a polymer is a polymer chain segment. The polymer chain segment has a highly non-linear force response to applied deformation. The strain energy of a single polymer chain is entropically based and must account for finite extensibility at the forces relevant to mechanophore activation. Following Kuhn, the polymer chain has initial end-to-end distance $l_0 = l_k\sqrt{N}$, where $l_k = l_c/N$ is the Kuhn segment length and N is the number of Kuhn segments along the contour length (l_c) of the polymer chain segment.[19] The strain energy Ψ is defined in terms of the stretch on the chain:

$$\psi = k_b \theta N\left(\frac{\lambda}{\sqrt{N}}\beta + \frac{\ln\beta}{\sinh\beta}\right) - \theta_c \tag{4.18}$$

where λ is the chain stretch, β is the inverse Langevin function, $\beta = L^{-1}\left(\frac{\lambda}{\sqrt{N}}\right)$ for the Langevin function defined as $L[\beta] = \coth\beta - \left(\frac{1}{\beta}\right)$, and c is a constant.[20] The force (f) on the chain is given by the derivative of the strain energy as:

$$f = \frac{l}{l_0}\frac{\partial\Psi}{\partial\lambda} = \frac{k_b\theta}{l_k} = L^{-1}\left(\frac{\lambda}{\sqrt{N}}\right) \tag{4.19}$$

N sets the locking stretch of the polymer chain; as the chain approaches this stretch, the force increases rapidly (see Figure 4.6).

The mechanophores can be one or more units on the polymer chain segment. Assuming equilibrium, the force across each unit of the polymer chain segment is equal and is equal to the overall polymer chain segment force (similar to how tension does not vary along a free standing rope). The polymer chain force is therefore the force acting on the mechanophores and can be used directly in eqn (4.13)–(4.17) to find the mechanophore rate of activation for a population of mechanophore-linked polymer chains subjected to identical force (for this perspective, amplification factors such as lever arms[21,22] in adjacent monomers are grouped into the mechanophore). The combination of the hyperelastic force-stretch response of the polymer chain and the exponential force dependence of the reaction rate results in mechanophore response that is highly sensitive to polymer chain stretch.

The polymer chain segments combine to form a network *via* chemical cross-links and/or physical entanglements. This network determines how deformation of the bulk polymer becomes force on the polymer chain segment and also determines the stress response of the bulk polymer. Network morphology is a critical theme for both the experimental and theoretical work performed on elastomers. For example, if the network is formed from chemical cross-links then some chain segments have to deform with the bulk deformation, and force is necessarily transmitted to the mechanophores embedded at the sites of molecular deformation. Conversely, in a physically entangled network, the polymer chains may slip past each other rather than stretch, resulting in minimal mechanophore response. Below, we discuss:

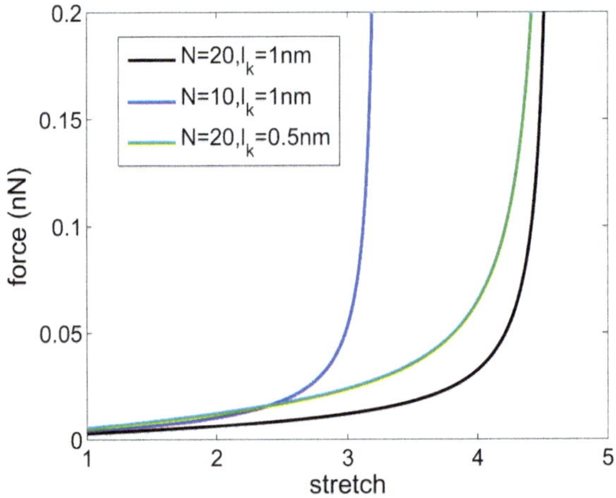

Figure 4.6 Polymer chain mechanical behavior. Characteristic force-stretch response of polymer chains according to finite extensible freely jointed chain model.[20] Locking stretch decreases with decreasing N. Stiffness increases with decreasing l_k.

first the existing experimental data on optically trackable MCR elastomers, and second the corresponding theory.

Representative stress and color change response for an SP-linked elastomer under monotonic tensile loading is shown in Figure 4.7. The network stress–strain behavior is hyperelastic. The color (here expressed as Red–Green–Blue ratios to avoid thickness effects) is initially constant and then increases steeply at a strain of 100% as the elastomer approaches its locking stretch. If this SP conversion were recorded as fluorescence or color intensity rather than color ratio, the recorded intensity would decrease at low strains, as the specimen thins (and number of mechanophores per area decreases) prior to achieving strains at which activation is observed. The onset of activation (labeled on Figure 4.7 as ε*) is defined as the point at which mechanophores

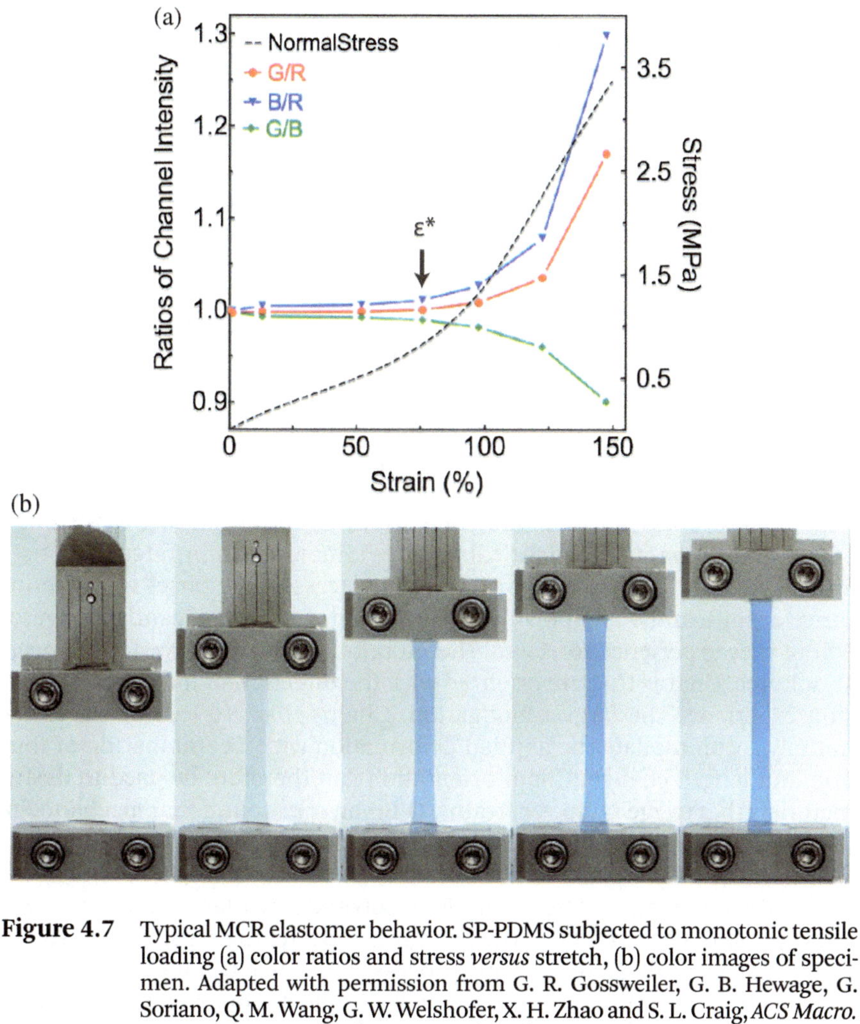

Figure 4.7 Typical MCR elastomer behavior. SP-PDMS subjected to monotonic tensile loading (a) color ratios and stress *versus* stretch, (b) color images of specimen. Adapted with permission from G. R. Gossweiler, G. B. Hewage, G. Soriano, Q. M. Wang, G. W. Welshofer, X. H. Zhao and S. L. Craig, *ACS Macro. Lett.*, 2014, **3**, 216–219.[8] Copyright (2014) American Chemical Society.

are clearly being mechanically triggered and is convenient for comparing among experiments. As should be clear from our discussion of mechanophore kinetics, a non-zero quantity of mechanophores are being mechanically triggered as soon as any stress is applied to the polymer, so this onset of activation is a matter of experimental practicality rather than something with a strict physical definition. Typically, the onset of activation is defined by an intensity slope change (*e.g.* ref. 23) or, for SP only, the intensity relative to a UV-activated specimen intensity (*e.g.* ref. 24). In a monotonic tensile test to failure of an elastomer, signal intensity typically increases monotonically through failure following the onset of activation. A summary of elastomer experiments and results is given in Table 4.1.

The potential requirement for chain orientation is a central theme for triggering mechanophores within elastomers. Beiermann *et al.* performed polarized fluorescence measurements on SP-linked poly(methyl acrylate) (PMA) at discrete levels of stretch under uniaxial tensile loading, comparing orientation of mechanically activated mechanophores and UV-activated mechanophores.[25] They found that the mechanically activated mechanophores are highly aligned with the direction of applied tension at the onset of activation and that orientation decreases slightly as stretch increases. In contrast, the UV-activated mechanophores (representative of all the mechanophores in the material) start out randomly oriented and monotonically increase in orientation (see Figure 4.8a). In a follow-up study, the authors captured birefringence at the same time as fluorescence intensity during continuous uniaxial tensile loading of the same SP-linked linear PMA material.[17] Birefringence measures the overall polymer chain orientation. Since the mechanophores are in the polymer backbone it is reasonable to assume that the mechanophore orientation matches the polymer chain orientation. Fluorescence clearly starts to increase significantly only after the polymer chains are mostly aligned (see Figure 4.8b). This is often a highly rate-dependent process, and the birefringence value corresponding to the onset of activation was nearly identical at three different strain rates (0.004 s^{-1}, 0.02 s^{-1}, 0.10 s^{-1}), whereas the stretch and stress at the onset of activation varied significantly. The polymer chains need to be significantly elongated before they experience force and therefore transmit that force to the mechanophores. Chains that are oriented with the direction of macroscale extension experience the largest elongation. Chains that are less well oriented initially accommodate the applied deformation with a combination of rotation and length change. Chain orientation can therefore be used to design materials that trigger at lower strains. O'Bryan *et al.* found that mechanochromism only occurred in SP-linked polycaprolactone (PCL) prior to fracture if the specimens were stretched at elevated temperatures as part of the initial material processing.[7] Interestingly, Peterson and coworkers found activation in 3D printed SP-linked PCL without any pre-stretching step, a result that is likely due to the extrusion printing process, which itself should induce chain alignment.[26,27] Ramirez *et al.* prepared a triblock copolymer of PS-PB-PS (polystyrene – polybutadiene – polystyrene) with the length extending

Table 4.1 Key experimentally determined parameters for MCR elastomers tested under uniaxial tension. Mechanophore indicated before host polymer. Most values approximated from published plots. "–" indicates that information was not found in the publication.

Material	Elastic modulus (MPa)	Failure stretch	Failure stress (MPa)	Activation stretch	Activation stress (MPa)
SP:					
Linear PMA[5,17,25]	2	10	30	4	5
Filled PDMS[8,29,30]	1	3	5	1.8	0.8
PU[33]	40	–	–	1.2	10
Segmented PU 22 wt% hard[16]	5	6	15	1.5	6.7
Segmented PU 40 wt% hard[16]	15	6	30	1.2	17
Segmented PU - UPy[36]	8.4	11	22	6.5	11
Segmented PU - BDO[36]	9.1	8	10	5.5	6.9
PS_{15}-$PnBA_5$-PS_{15}[38]	10.7	5.3	3	3	2.5
PS_{15}-$PnBA_3$-PS_{15}[38]	0.32	>11	0.25	7	0.15
PS_{19}-$PnBA_4$-PS_{19}[38]	0.45	>11	0.25	None	None
PCL[26,27]	300	8	37	<8	<37
PCL[7]	–	2	–	1.2	–
Segmented PU-UPy[23]	2	9	27	5	5
Segmented PU-UPy M_w10 kDa[37]	2.6	11	46	6	12
Segmented PU-UPy M_w20 kDa[37]	4.1	11	42	6	10
Segmented PU M_w10 kDa[37]	2.1	10	6.5	7.5	3
Segmented PU M_w 20 kDa[37]	2.3	10	6.5	7.5	3.5
DABBF:					
Segmented PU[14]	2	15	6	1.5	1.5
gDBC:					
PB[32]	0.1	22	0.2	None	None
PS-PB-PS film[69]	0.1	–	20	–	<20
PS-PB-PS fibers[69]	–	–	36	–	<36
Dioxetane:					
Linear PMA[11]	1.5	3.8	35	3	5
Cross-linked PMA[11]	2.5	4.2	45	4	10
Segmented PU M_n 43 kDa[35]	12	3.5	90	3.1	40
Segmented PU M_n 38 kDa[35]	11	3.4	50	2.5	20
Segmented PU M_n 12 kDa[35]	8	2.4	10	1.9	3
Segmented PU/plain PU[35]	20	3.2	60	<3.2	<60
PE/plain PU[35]	20	3	50	<3	<50
PA/plain PU[35]	20	3.2	70	<3.2	<70
DN[34]	0.8	3.3	6.1	2.5	1.6
TN[34]	1.5	3.8	6.8	1.5	1.7

mechanophore *g*DBC (*gem*-dibromocyclopropane) covalently linked within the PB segment both as an unoriented film, produced by solvent casting, and as oriented fibers produced by electrospinning.[28] They found that the aligned electrospun fiber mats had seven times more mechanophore activation than the randomly oriented polymer film after testing to specimen failure under tension.

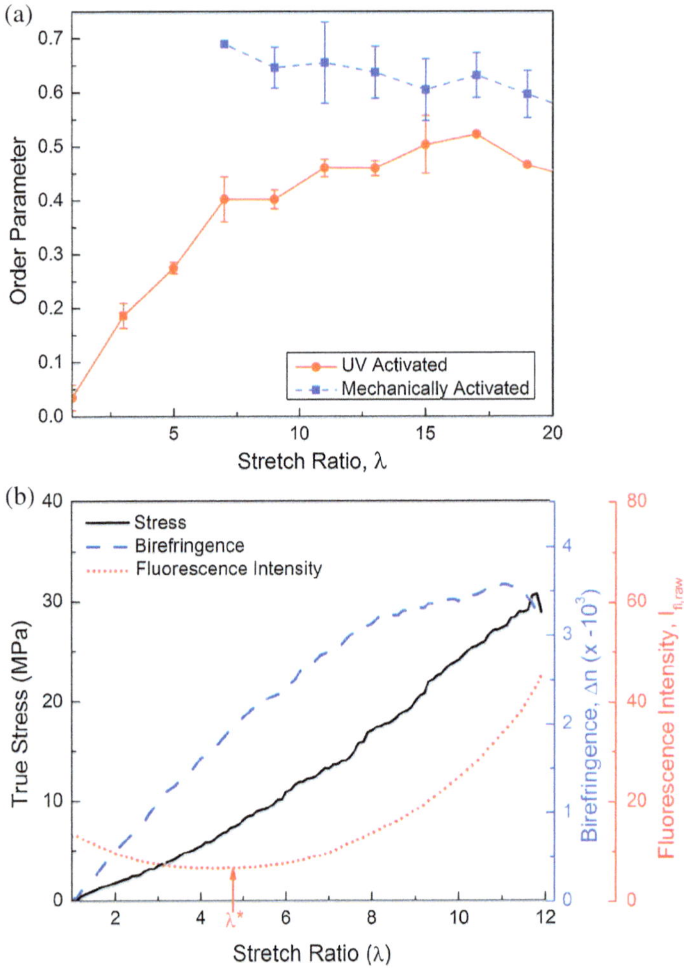

Figure 4.8 Results of SP-linked linear PMA subjected to monotonic uniaxial tensile loading. (a) Order parameter for mechanically activated mechanophores compared to UV-activated mechanophores, reprinted with permission from B. A. Beiermann, S. L. B. Kramer, J. S. Moore, S. R. White and N. R. Sottos, *ACS Macro. Lett.*, 2012, **1**, 163–166.[25] Copyright (2011) American Chemical Society. (b) Stress, fluorescence intensity and polymer chain orientation (given by birefringence) as a function of stretch. Reprinted from ref. 17 with permission from John Wiley and Sons. Copyright © 2014 WILEY-VCH Verlag GmbH & Co. KGaA, Weinheim.

While a single polymer chain will only activate under tension, mechanophores within bulk polymers can be activated under tension, compression, shear or combined loading – each of which generates localized tension at the molecular level. However, relatively few experiments have been performed to determine the stress state dependence of activation. The most complete stress state picture for a single elastomer–mechanophore combination exists for SP-linked PDMS (polydimethylsiloxane) performed by the Zhao and Craig groups.[8,29,30] This PDMS is a filled commercial elastomer called Sylgard 184 that is moderately stiff, highly stretchable and minimally rate-dependent – all desirable features for assessing the role of stress state on mechanochemistry. Using thin films, the authors conducted uniaxial tension, uniaxial strain and equibiaxial strain experiments. They found that in all cases color change occurred. Further, the others found a direct correlation between the first invariant of the Cauchy–Green deformation tensor and fluorescence intensity.[29] Physically, the first invariant represents the square of an averaged network stretch. Most rubber elasticity models for strain energy and stress are functions of the first invariant.[31] The authors then validated the first invariant correlation with uniaxial strain applied to a specimen with a hole and a specimen with an edge crack and found reasonably good agreement.[30] The authors also performed compression experiments on spherical specimens, the mechanochemical reaction clearly occurred, but the extent is difficult to compare to the thin film results due to specimen thickness difference and likely residual solvent.[18] In a different set of work, the Craig group showed that gDBC-linked PB could be activated to measurable quantities (0.3% from a single compression, ~2% from multiple) under compression but not under tension.[32] Although the authors do not report precise strain values, it is clear from pre and post pictures of the specimens that the compression specimen was tested to much larger true strain. The polymer fails under tension at a stress of only 0.2 MPa, whereas compression stresses are in the 10s and 100s of MPa. Quite aside from how stress state at a material point is distributed as force to the mechanophores, activation can be limited by structural failure associated with different loading modes. Even after significant polymer chain damage, compressive loading continues to allow for transmission of force to the polymer chains, whereas specimens lose the connectivity to transmit force after moderate damage in either tensile or shear loading.

MCR elastomer response is strain-history- and time-dependent. This dependence arises both from the elastomer mechanical dependence on strain history and time (varies among elastomers) and the mechanophore population kinetics (varies among mechanophores and with applied molecular force). To isolate mechanophore population kinetics, Lee *et al.* conducted stress relaxation experiments on SP-linked PU (polyurethane) in which the SP was optically driven to the closed state after stress reached a relatively steady state and then allowed to reach dark equilibrium. This experiment showed that the force-biased equilibrium is reached in one to two hours at a stress of around 25 MPa.[33] Conversely, Lenhardt *et al.* found that activation of *g*DBC functionalized PB did not depend on time for a range of 15 s to 600 s under controlled pressure within a pellet press.[32] The lack of time dependence here likely reflects relaxation in the PB matrix (*i.e.* the applied stress dissipates in the first 15 s), but secondary effects could in theory play a role; *g*DBC is present in the polymer at much higher percentages than SP and activation is therefore much more likely to influence (in this case reduce) subsequent activation. For linear PMA, which is an elastomer whose network is formed solely by physical entanglements, both the mechanical and mechanochemical response of SP-linked PMA were found to be strain rate dependent.[17] In monotonic testing at three different strain rates it was seen that the activation stretch decreased and the corresponding stress increased with increasing strain rate, but that polymer chain orientation was relatively unchanged. Further, in stress relaxation testing at a stretch beyond the activation stretch, it was found that fluorescence continued to increase even as stress decreased, reaching a roughly steady state value over 100 s. This continued increase could be in part an effect of stress redistribution throughout the polymer, but most likely can be primarily attributed to the activation not having time to reach equilibrium during the initial loading portion. The linear PMA material system highlights the often occurring interplay between force and the time over which that force is applied to determine degree of activation. Faster testing rates correspond to higher stresses (and corresponding greater stretch of the polymer chain), but faster testing also means less time at a given macroscale strain. Similarly Zhang *et al.* characterized a viscoelastic polyurethane at different strain rates and found that stiffness and strength increased and activation stretch decreased with increasing strain rate.[23] Chen *et al.* tested dioxetane in both cross-linked PMA and linear PMA under monotonic tensile loading.[11] Emission in the cross-linked polymer started at a similar strain to the linear polymer but occurred in a more localized region around which the fracture eventually occurs (the intensity values cannot be directly compared because a different acceptor is used to modify the emission color). Ducrot *et al.* synthesized a multinetwork elastomer for which each interpenetrating network has equivalent chemical structure and cross-linking, but different levels of built-in pre-strain.[34] Dioxetane was covalently incorporated into the network with the largest pre-strain so that on average they would break well before the other bonds. The mechanophores emitted light starting from a stretch of 2.5 for the double network and 1.5 for the triple network (larger pre-strain). Under an increasing cyclic loading history with three cycles to each stretch, light was emitted primarily when a new stretch level was reached, with a small and decreasing amount of light emission during second and third reloading to a given stretch. These results indicate that even in this highly targeted polymer architecture not all mechanophores activate when a particular force is first reached, reiterating the central role of mechanophore kinetics and/or contributions from plastic deformation or similar material relaxations. The reduction in activation with subsequent cycles to the same stretch is probably to a large extent the result of damage to the highest pre-strained network.

Mechanophore response can be tailored by creating an intentionally inhomogeneous distribution of strain on the nanoscale – naturally this will change the mechanical behavior of the polymer as well. Segmented polyurethanes and triblock copolymers are two motifs that can result in strain being distributed unevenly to different portions of the microstructure. When a mechanophore is embedded in the compliant segment it will follow elastomer physics, but the strain within that segment should be larger than the global strain.

Lee *et al.* incorporated SP within the soft phase of segmented polyurethane for two different ratios of hard segment (22 wt% and 40 wt%).[16] The material with the higher weight percent of hard segment has more phase separation and a stiffer and stronger mechanical response than with a lower percent of hard segment. Fluorescence resulting from mechanophore activation for the 40 wt% hard segment PU increases faster than the 22 wt% hard segment PU at low stretches and then saturates at large stretch. Chen *et al.* found that the mechanoluminescence of segmented PU-dioxetane was six times that of the equivalent PMA-dioxetane system.[35] In a second part of the study, they compared the mechanoluminescence of dioxetane-linked PE (polyester), PA (polyamide) and PU mixed (non-covalently) with PU without mechanophores. While the three materials exhibited similar overall mechanical properties, the mechanoluminescence depended strongly on the interaction strength of the dioxetane-containing component with the host PU polymer. The mechanoluminescence decreased from PU to PA to PE, in accordance with decreasing strength of that interaction. This mixing approach showed the importance of transmitting strain to the mechanophore-linked component. Fang *et al.* compared SP and UPy (ureidopyrimidinone) in a polymer with SP and a covalent cross-linker in an equivalent polymer under tensile loading.[36] The UPy units form hydrogen bonded hard domains. Since the UPy are in general more weakly bonded than SP, the UPy hard domains fragment before the SP activates under monotonic tensile loading. As expected, the SP activates earlier in the covalently bonded system than in the hydrogen bonded system (stretch of 5.5 *versus* 6.5) since fragmentation of the hydrogen bonded domains releases extra contour length within the polymer. In a follow-up study, Chen *et al.* compared SP-linked pTHF (polytetrahydrofuran) capped with UPy to the same polymers capped with non-hydrogen bonding urethanes.[37] While both types of materials exhibited color change, the UPy-capped polymers (higher stiffness and strength, larger failure strain) activated earlier and more strongly as a function of strain for two significantly different strain rates (0.0017 s^{-1}, 0.05 s^{-1}). According to X-ray scattering profiles taken at different strain levels for each polymer architecture, the UPy-capped polymers have greater initial phase segregation, greater orientation of the soft segment during deformation and strain induced crystallization parallel to the tensile axis. Imato *et al.* linked DABBF into the soft segment of a segmented PU.[14] Activation increases linearly with strain starting at 50% under monotonic loading. When held at a constant strain of 50% after prior stretching to 300%, the color from the cleaved DABBF units slowly dissipates to the room temperature equilibrium value over 5 hours, the authors suggest that the chain mobility in the soft segment allows the cleaved DAABF units to recombine from any pair of active mechanophore halves. Jiang *et al.* covalently bonded SP at the center of a PS-PnBA-PS (polystyrene – poly(*n*-butylacrylate) – polystyrene) triblock copolymer where PS is the stiffer component.[38] Higher ratios of PS:PnBA resulted in higher stiffness and strength and lower ductility. As long as the overall material remained elastomeric, the mechanophore activated prior to failure; with higher ratios of PS to PB, the color change was more intense. Similarly, Ramirez *et al.* showed that activation of triblock copolymer film of gDBC in PS-PB-PS was greater than that of gDBC in pure PB.[28]

As is evident from the body of experimental results presented above, any theoretical basis for connecting macroscale loading to mechanochemical activation within an MCR elastomer must consider the polymer microstructure. For elastomers the key microstructural aspects are the cross-link density and distance between them, strain distribution on the nanoscale and initial orientation distribution of the mechanophore-containing polymer chains.

The first theory to make a macroscale to mechanophore connection for elastomers was published in 2014 by Silberstein and coworkers.[39] The theory is for an initially homogeneous isotropic elastomer with physical cross-links. Both the orientation distribution and the length between cross-links are initially uniform with no particular molecular features to create strain localizations. A key self-consistent component of the theory is that the forces on the polymer chains produce both the stress and the mechanophore response. The theory was validated with respect to SP-linked linear PMA data from the Sottos group.[17] Each polymer chain segment was assumed to follow Langevin chain statistics as previously discussed and described by eqn (4.18) and (4.19). Given the small quantity of SP, it was assumed that activation does not noticeably increase polymer chain segment length. The polymer network is numerically approximated by a discretization into 21 distinct populations of representative chains distributed isotropically about a unit sphere in the reference configuration. This approach was selected over the more well-known eight-chain or full chain models for its explicit representation of polymer chain force as a function of orientation. The total strain energy density is the weighted average of the strain energy of each of the populations:

$$\psi = \sum_{i=1}^{21} w^i n^i \Psi^i \tag{4.20}$$

where w^i is the weighting factor for population i, $w^i n^i$ is the number density of population i and Ψ^i is the strain energy of population i. Because the network is physically (rather than chemically) cross-linked, the number of Kuhn segments between cross-links will evolve within the constraint that the total number of Kuhn segments does not change. The network stress is then derived from the network strain energy as:

$$\mathbf{T} = \frac{1}{J}\left[\sum_{i=1}^{21} w^i n^i f^i l_\mathrm{k} \sqrt{N^i} \frac{\partial \lambda^i}{\partial \mathbf{F}}\right]\mathbf{F}^\mathrm{T} \tag{4.21}$$

Activation occurs within each of these discretized elements of the polymer network according to its particular chain force input into eqn (4.13)–(4.17) as described in Section 4.3. With no fitting parameters to the activation model, both the highly non-linear nature and proper rate dependence of the activation response are predicted, but activation occurs at larger strains in the simulation than seen experimentally (see Figure 4.9). This delay in activation could be due to uncertainty in the *ab-initio* calculations used to determine the force-modified energy barrier, short time force fluctuations from the internal chain motion and/or initial entanglement length inhomogeneity.

The Zhao group improved upon this theory by adding in the critical effect of entanglement length distribution.[30] This model was validated with respect to SP-linked PDMS. Because PDMS is a chemically cross-linked material, viscous

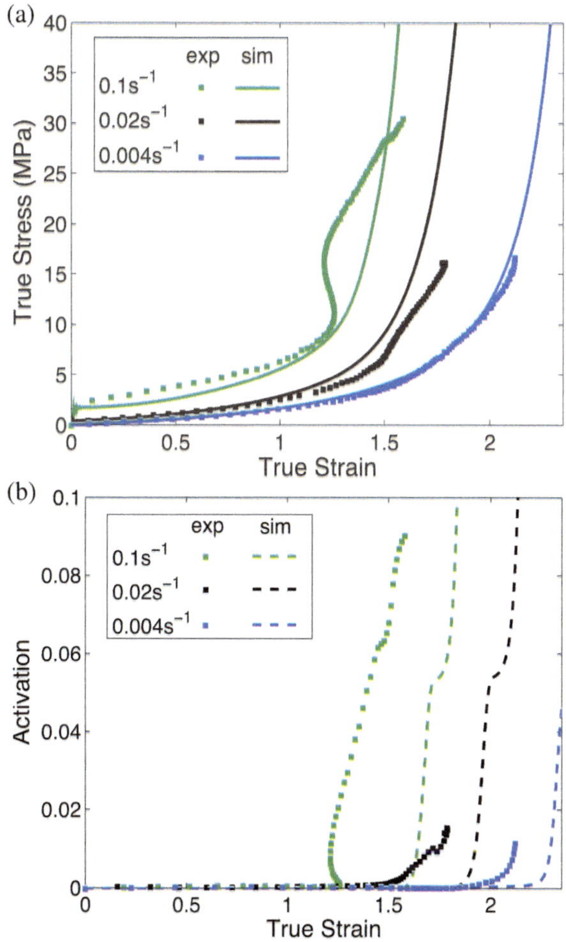

Figure 4.9 Experimental and simulated response of SP-linked PMA under monotonic loading at three strain rates. (a) Stress *versus* strain. (b) Activation *versus* strain. Adapted from *Journal of Mechanics and Physics of Solids*, 63, M. N. Silberstein *et al.*, Modeling mechanophore activation within a viscous rubbery network, 141–153. Copyright 2015, with permission from Elsevier.[39]

slip of the polymer chains was not needed. These authors chose to adopt the incompressible version of the computationally facile eight-chain model for the polymer network. In the eight-chain network, there is a single effective polymer chain stretch (λ_{chain}) that is directly related to the first invariant.

$$\lambda_{\text{chain}} = \sqrt{\frac{I_1}{3}} \tag{4.22}$$

The polymer network is formed from the superposition of eight-chain networks with varied cross-link densities (N-values) with all networks following the same deformation gradient. The force of each interpenetrating network chain is again given by eqn (4.19), but with the effective chain stretch as the polymer chain stretch and using the number of Kuhn segments between cross-links for that particular network. The network stress is the derivative of the weighted strain energy of the superposed network as given by:

$$\mathbf{T} = \frac{k_b \theta}{3}\left[\sum_{i=1}^{m} \frac{n^i \sqrt{N^i}}{\lambda_{\text{chain}}} L^{-1}\left(\frac{\lambda_{\text{chain}}}{\sqrt{N^i}}\right)\mathbf{B}\right] - p^* \mathbf{1} \tag{4.23}$$

where m is the number of networks of different cross-link densities n^i, segment length between cross-links N^i and p^* is the pressure required to satisfy the boundary conditions for an incompressible material. The forces on the polymer chains are again the same forces that drive activation within that network as given by eqn (4.13)–(4.17). Activation has an independent tracked value within each of the interpenetrating networks. A central output of this work was showing that polymer networks of various distributions of length between cross-links can have nearly identical stress responses and drastically different mechanophore responses to applied deformation. Mechanochemical response can therefore be tailored with polydispersity. After fitting the activation parameters to uniaxial strain data, the authors showed a strong ability to predict fluorescence of the same PDMS film under several loading geometries with complex stress states.

The theory for MCR elastomers, while quite recent, is strong in both physical/mechanistic underpinnings and practical usability. One addition that could significantly improve the predictive capabilities of the theory would be to incorporate temporal inhomogeneity in the local polymer chain stretch. With the exponential dependence of activation kinetics on force, the bulk polymer is even more sensitive to temporal fluctuations than to spatial inhomogeneity; however, these fluctuations are difficult to quantify. Additionally, the generality of the theory could be validated by testing systems with mechanophores other than spiropyran. For example, the same polymer could be synthesized with spiropyran and gDBC. Two natural and fairly straightforward extensions of the theory would include gels and phase segmented polymers. The mechanochemistry framework can be readily incorporated into the existing polymer models for each.

4.6 Glassy Polymers

Glassy polymers behave distinctly different from elastomers both mechanically and mechanochemically. Glassy polymers are by definition polymers below their glass transition temperature (T_g). Mechanically, the elastic moduli are roughly one GPa, orders of magnitude stiffer than elastomers. Depending on the polymer chemical structure, thermal history, temperature relative to the T_g and loading mode, the polymer will either behave elastically and fail in a brittle manner, or yield and deform plastically. A typical glassy polymer stress–strain curve is shown in Figure 4.10 and essentially every feature is subject to the complexity of glassy behavior: the stress–strain curves are often non-monotonic, exhibiting first strain softening and then strain hardening; fracture can occur *via* shear or crazing; plasticity is both temperature- and rate-dependent; the yield stress decreases and ductility increases with increasing temperature and decreasing strain rate. Yielding in polymers is a gradual event starting from the apparently elastic regime and involving an evolution of the microstructure typically referred to in terms of either free volume or mobility. The fracture strain under tensile loading is much smaller than in elastomers, spanning from a few percent to 100%. Also, unlike elastomers, the deformation does not recover on a laboratory time frame unless the temperature is raised above the T_g.[40–44]

Relatively few experiments have been conducted on MCR glassy polymers. The vast majority of published mechanochemical testing of glassy polymers is on SP-linked poly(methyl methacrylate) (PMMA) as pioneered by the Sottos group.[4] These papers characterize the time, temperature, plasticization and loading mode dependence of activation as discussed below. We also include here related work on thermoplastic elastomers and triblock copolymers that were characterized below their T_g. The dominant theme for glassy polymer mechanochemistry is the coupling between activation and plastic flow.

Davis *et al.* first characterized the mechanochromic response of linear SP-PMMA by performing compression experiments on PMMA beads.[5]

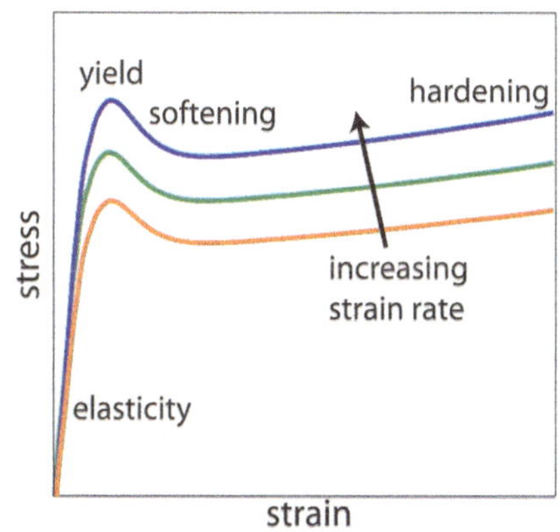

Figure 4.10 Sketch of typical glassy polymer behavior subjected to monotonic tensile loading.

This bead compression loading results in a complex stress state with tensile stress across the center of the equatorial plane. These experiments proved that glassy polymers could be activated. The color change occurred after the yield stress and was concentrated in the portion of the specimen subjected to tension. Beiermann *et al.* tested the same polymer under pure tensile loading.[25,45] The dogbone specimens were imaged after monotonic tensile loading to failure at temperatures from 22 °C to 120 °C (T_g of PMMA is 127 °C).[45] Color change resulting from SP activation was largest for the test conducted at 90 °C, which seemed to have the best combination of stress and ductility for mechanophore activation. The dogbones were also imaged during monotonic tensile loading at room temperature but plasticized with a solvent, effectively bringing the polymer closer to T_g. The fluorescence is flat through yield and then increases monotonically until failure (see Figure 4.11). The correlation of mechanophore orientation and activation was also determined during uniaxial tensile loading.[25] The orientation parameter for

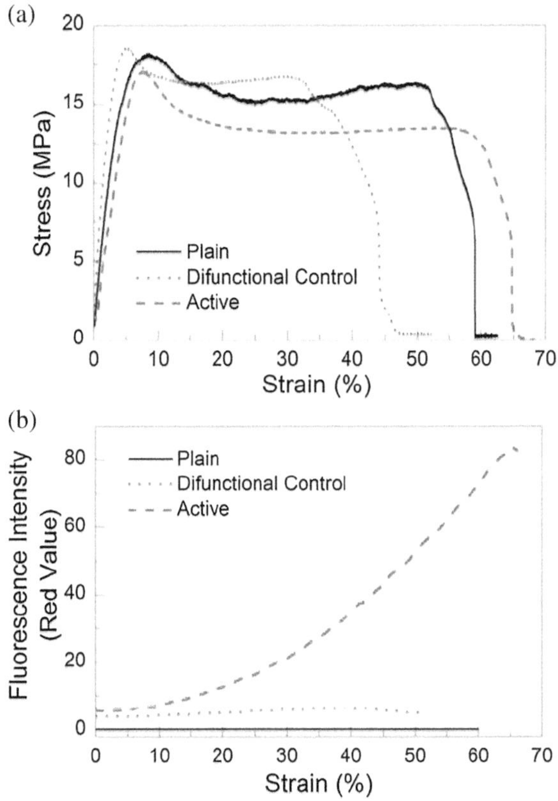

Figure 4.11 Results of SP-linked linear PMMA subjected to uniaxial tensile loading while plasticized with methanol. (a) Stress *versus* strain. (b) Fluorescence intensity *versus* strain. Reproduced from ref. 45 with permission from the Royal Society of Chemistry.

activated mechanophores is moderately higher than for the polymer overall, and less than for PMA when activation initiates.

Kingsbury *et al.* characterized the fluorescence of lightly cross-linked PMMA under torsional loading.[24] Three different cross-linkers were used: EGDMA (ethylene glycol dimethacrylate), PEG (poly(ethylene glycol) dimethacrylate) 550 and PEG 750 increasing in length from 8 Å to 37 Å to 50 Å. All three versions of the polymer activate after the yield peak. The strength of the polymer increases with decreasing cross-link length and increasing strain rate. The strains at the onset of activation are given in Table 4.2. At slow rates activation strain is insensitive to the difference in cross-link length, at fast rates the polymer with the longer cross-linker activates later. In a follow up study, Degen *et al.* subjected the EGDMA cross-linked SP-PMMA to creep loading.[46] Activation was found to occur at stress levels below the monotonic activation stress. Further, the time to activation (including both monotonic and creep data) correlated strongly with the maximum strain rate, and time to activation correlated one-to-one with time to maximum creep rate (see Figure 4.12). Basically, activation under creep loading occurs after the material moves through the creep microstructural equivalent of the monotonic yield peak. Kim *et al.* took this same EGDMA cross-linked SP-PMMA and monitored the mechanoactivation *in situ* during tension and compression at different strain rates and temperatures.[47] Activation under tension and compression was found to be quite similar with the later onset of activation in compression corresponding to the delayed yield peak. Additionally, at temperatures approaching PMMA's T_g, the authors suggests that increasing molecular mobility in the polymer is responsible for some activation occurring prior to the polymer progressing through the yield peak. Lee *et al.* induced activation in EGDMA cross-linked SP-PMMA by solvent swelling.[48] The fluorescence intensity was found to depend on mass uptake and identity of the solvent. Both the mass uptake and fluorescence reduced when the percent of cross-linker was increased (see Figure 4.13). In this study, the solvent provided both the molecular level force and the mobility required for glassy polymers to activate.

Hemmer *et al.* characterized the fluorescence response of SP-linked lightly cross-linked PMMA under compression at a strain rate of approximately 10^3 s^{-1} using a split-Hopkinson bar.[49] In order to isolate the pure mechanical activation from the thermal activation resulting from the mechanically induced temperature rise, the same polymer with a difunctional control version of

Table 4.2 Strain at onset of activation for cross-linked SP-PMMA under monotonic torsional loading as a function of strain rate and cross-linker length.[24]

Shear strain rate	PMMA-EGDMA	PMMA-PEG 550	PMMA-PEG 750
0.0001 s^{-1}	0.45	0.46	0.45
0.001 s^{-1}	0.46	0.45	0.67
0.01 s^{-1}	0.48	0.57	0.77

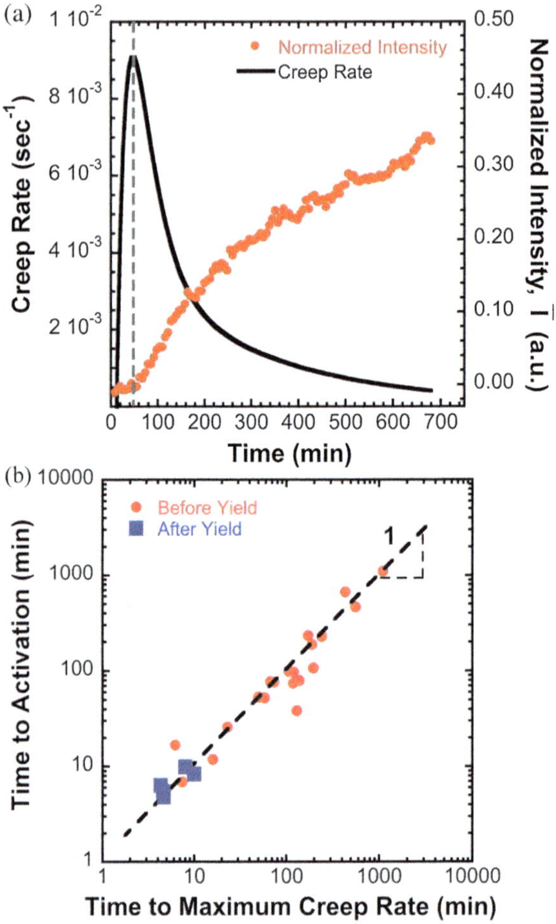

Figure 4.12 Activation trends for EGDMA cross-linked SP-PMMA tested under monotonic and creep torsional loading. (a) Creep rate and fluorescence intensity *versus* time. (b) Comparison of time to activation with time to maximum creep rate showing a one-to-one correlation. Reprinted with permission from C. M. Degen, P. A. May, J. S. Moore, S. R. White and N. R. Sottos, *Macromolecules*, 2013, **46**, 8917–8921.[46] Copyright (2013) American Chemical Society.

SP was also tested. Optical characterization was only performed on the shattered pieces after testing. The color change was found to be heterogeneous ranging from no change to deep purple. Fracture surfaces exhibited the greatest fluorescence intensity. In general, activation of the active polymer and the control polymer were similar; however, there were fibrillar regions likely associated with crazing that did have significantly higher fluorescence in the active polymer. These data further supports the concept that plasticity is needed for mechanochemical activation in glassy polymers regardless of the magnitude of the stress.

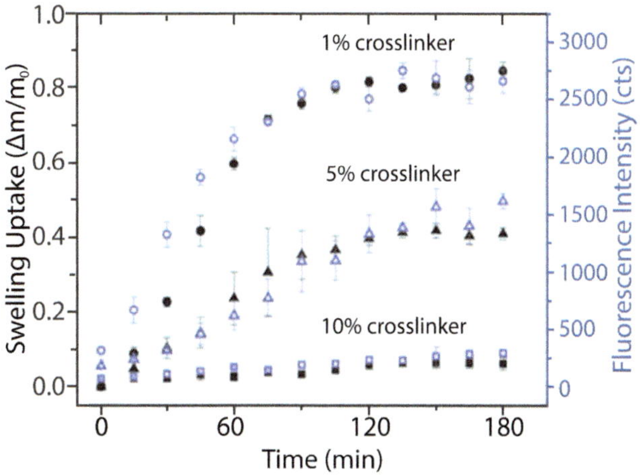

Figure 4.13 Swelling uptake and fluorescence intensity *versus* time for EGDMA cross-linked SP-PMMA activated by solvent swelling. Adapted with permission from C. K. Lee, C. E. Diesendruck, E. J. Lu, A. N. Pickett, P. A. May, J. S. Moore and P. V. Braun, *Macromolecules*, 2014, **47**, 2690–2694.[48] Copyright (2014) American Chemical Society.

Celestine *et al.* characterized fluorescence induced by mode I crack propagation in rubber toughened EGDMA cross-linked SP-PMMA.[50] Specimens were pre-stretched under uniaxial tension to a strain of 0.35, illuminated with light to drive all MC back to the SP state, notched on one side and then extended to failure. This notching and extension is referred to as a single edged notch test (SENT). They found that pre-stretching, which aligns the polymer chains and effectively erases the polymer annealing history by traveling through yield, increases the stress and fluorescence during the subsequent SENT deformation (see Figure 4.14a). Further, the authors compared the fluorescent area at the crack tip with the theoretically calculated plastic zone and found a near 1:1 relation within the limits of the specimen dimensions (see Figure 4.14b).

Several other polymers were tested under uniaxial tension in their glassy state. Nevertheless, none showed significant activation. Grady *et al.* tested SP-linked PS and found that the material failed in a brittle manner at a strain of 0.08 and showed no color change or increased fluorescence.[51] Interestingly this same SP-PS did change color when shock loaded as a thin film, the film also showed significant plastic deformation. Jiang *et al.* synthesized a SP-linked glassy triblock copolymer of PS-PnBA-PS with a PS:PnBA ratio of 9:15. The material started to fracture right around the yield strain of 20% and only showed color change at the fracture surface, which clearly underwent plastic deformation.[38] Chen *et al.* characterized a dioxetane-linked PMA below its T_g and a dioxetane linked PMMA and found that each failed in a brittle manner and did not emit light.[11]

For glassy polymers, unlike elastomers, there is no analytical theory to relate macroscale stress to molecular level force. Molecular dynamics is the

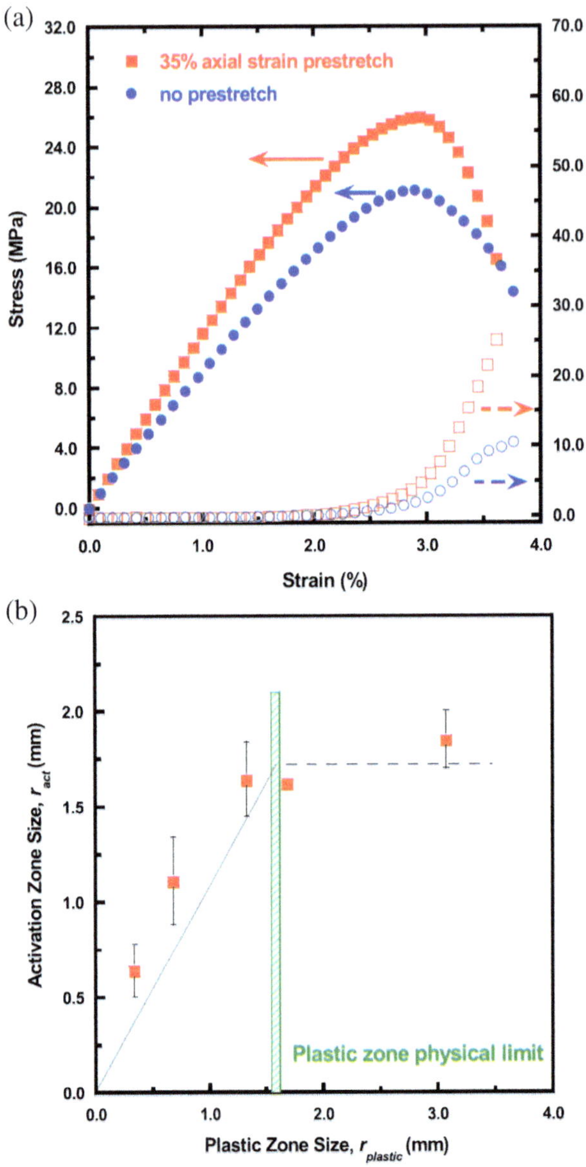

Figure 4.14 Results of pre-stretched rubber toughened EGDMA cross-linked SP-PMMA subjected to SENT testing. (a) Stress and fluorescence *versus* strain. (b) Comparison of activation zone based on fluorescence intensity to theoretically predicted plastic zone around the crack tip. Reprinted from *Polymer*, 55, A.-D. N. Celestine *et al.*, Fracture-induced activation in mechanophore-linked, rubber toughened PMMA, 4164–4171. Copyright 2014, with permission from Elsevier.[50]

method of choice for investigating the relationship between polymer chemistry, polymer structure and mechanical properties.[40,41] It is well established from these studies that polymers are heterogeneous at atomistic length scales.[52,53] While molecular dynamic studies have addressed the question of how structure evolves with mechanical deformation and what the corresponding stress is,[40,41,54] the question of what magnitude of force is being placed on a bond or set of bonds within the bulk prior to failure is much more difficult to address. This atomic level force, however, is the force that is pertinent for mechanophore activation.

Silberstein *et al.* published the first continuum theory for mechanophore activation within a glassy polymer.[55] This theory is built on glassy polymer mechanics and the key findings from glassy polymer mechanochemistry data available at the time – chiefly, that activation does not occur until after the monotonic yield peak and that activation rate increases with increasing stress. The theory was specified to lightly cross-linked PMMA and validated with experimental data from the Sottos group.[24,46] The polymer mechanical behavior is captured with a large deformation elastic-viscoplastic polymer model. Critically, the model accurately captures the evolution of the material through the yield peak and post yield strain hardening (see Figure 4.15a,b). As is typical for polymer models, plastic deformation is driven by the deviatoric stress. Molecular dynamics simulations were used to quantify the relationship between macroscopic stress and local force on the mechanophore. Specifically, EGDMA cross-linked PMMA was simulated under monotonic shear deformation while both the system stress and the force on the cross-links were monitored. The remarkably simple result that post-yield force is directly proportional to shear stress was found and this constant of proportionality was then used to calculate force on the mechanophore from continuum shear stress. There are three key relations for determining the rate of mechanophore activation in this model. (1) Force on the mechanophore is directly related to the magnitude of the deviatoric stress from the continuum mechanical model:

$$\bar{f} = C\tau \tag{4.24}$$

where $\bar{f}$ is the mean force driving activation, C is the constant determined from the MD simulations and $\tau = \sqrt{\frac{1}{2}\mathbf{T}'\mathbf{T}'}$ is the scalar equivalent shear stress. The effective shear stress is used (rather than the total stress magnitude) in analogy to polymer plasticity modeling with the intent that this model holds for all modes of loading. (2) There is a distribution about this mean force that reflects both the spatial and temporal inhomogeneity present in glassy polymers. The distribution of forces is captured as a normal distribution with the standard deviation taken as a fitting parameter. (3) Activation is only permissible if there is sufficient molecular mobility (see eqn (4.25)).

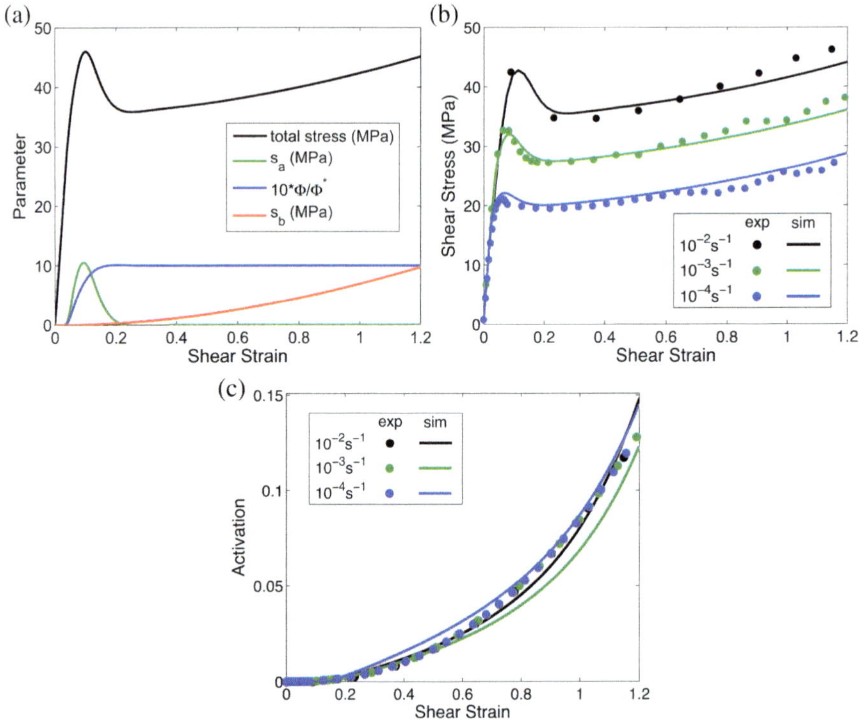

Figure 4.15 Simulation results for glassy polymer mechanochemistry model validated with EGDMA cross-linked SP-PMMA under monotonic shear loading. (a) Evolution of internal material parameters that control yield (s_a, ψ) and post yield strain hardening (s_b). (b) Stress *versus* strain at three different strain rates. (c) Activation *versus* strain at three different strain rates. Reprinted with permission from ref. 55. Copyright 2013, AIP Publishing LLC.

$$\frac{\phi}{\phi^*} > 0.99 \tag{4.25}$$

where ϕ and ϕ^* are the internal state variables indicative of molecular mobility used in the mechanical plasticity model to capture the yield peak evolution. Eqn (4.25) is a direct expression of the constraint that the rigid matrix imposes on the mechanophore kinetics determined computationally[56–58] or experimentally determined in a single molecule force pulling experiments.[9,58,59] The force is utilized in eqn (4.13)–(4.17), as described in Section 4.3. With only one fitting parameter (the standard deviation of the force), the model was able to predict activation as a function of strain rate for monotonic loading and as a function of shear stress for creep loading (see Figure 4.15c). However, the experimental results of Kim *et al.* published subsequently have revealed that a cutoff condition on ϕ/ϕ^* is overly simplistic and inaccurate for temperatures approaching T_g.[40] Clearly, a more sophisticated representation of the matrix inhibition of mechanophore kinetics is needed that accounts for both plastic flow and temperature relative to T_g. Further, the comparison of the onset of activation in tension and compression in Kim *et al.* suggests that activation in glassy polymers might be more similar to elastomers than previously assumed. In the polymer mechanics literature, the strain-hardening regime of glassy polymers is sometimes modeled with the same equations as rubber elasticity. However, the physical appropriateness of such a model is debated.

4.7 Composites and Coatings

Composites and coatings present a promising future direction for utilizing mechanochemical functionality. A composite with two components has three options for mechanophore placement: (i) within the first material; (ii) within the second material; and (iii) at the interface of the two materials. For a deformation tracking, damage warning or self-repairing system, the best location choice will likely depend at least in part on the expected mode of deformation and failure. To assess modes of damage, mechanophores with different signals could be placed at each of these possible damage locations. The predominant modes of particle-filled composite damage are debonding of the filler particle, cavitation of the filler particle or the matrix, and crack propagation through the matrix which may either be pinned by or deflected by the particle.[60–62] Coatings can fail from cracks within the coating, debonding of the coating from the underlying material or often a combination of both.[63]

There has been minimal experimental work on MCR composites. The work that has been published on particle-filled polymers placed the mechanophore in the matrix and did not contain a comparison to alternative mechanophore placements.[8,29,30,50] We suspect that the lack of data for the response of material systems with mechanophores placed at interfaces is due to the combined difficulty of synthesis and experimental validation rather than a lack of viability. Damage sensing at the fiber–matrix interface has been demonstrated using fluorescent proteins.[64,65] These fluorescent proteins were attached to the fibers, the conformational change upon detachment of the fibers from the matrix caused the fluorescence to disappear if the protein attachment to the fiber was strong enough. These proteins were shown to be effective for detecting micrometer-scale damage in both glass and carbon-fiber-reinforced epoxy.

Mechanophore triggering at an interface is fundamentally different from mechanophore triggering within the bulk. At an interface, the mechanophore attachment points either move with the ends of the interface or they detach from the interface. Conceptually, the interface is a more extreme version of a segmented polymer with the strain localized within one polymer segment. Should the mechanophore become detached, then it becomes analogous to an uncross-linked chain with friction, topology and rate of deformation critical to the strain and force sustained across the mechanophore.

Sundaram and Silberstein recently published a first study on mechanophore triggering at a sparsely populated interface subjected to shear loading.[66] Both a kinematic and a molecular dynamics model were developed to investigate the basic mechanisms and tradeoffs for solid interface mechanochemistry. In each approach the mechanophore was modeled as two beads bonded by a double well potential. The double well potential captures the basic physics seen across mechanophores: with applied force the energy barrier for the forward reaction is lowered and the energy barrier for the reverse reaction is increased. The beads were then each bonded to a different substrate by a Morse potential which is harmonic for small displacement and can debond when subjected to sufficient force. The first essential requirement for activation is that there be sufficient relative displacement of the two sides of the interface (see Figure 4.16a). The minimum relative motion for activation to initiate is roughly the relative length of the two equilibrium mechanophore configurations. The second requirement for activation is that the bond that attaches the mechanophore to the substrate be sufficiently stronger than the bond between the two mechanophore beads (analogous to what was seen experimentally in the fluorescent protein fiber-epoxy system). In the athermal case, with only bonded interactions, there is a sharp cutoff set by comparing the maximum force supported by the two different covalent bonds; when there is significant thermal energy or non-bonded interactions this cutoff is somewhat blurred (see Figure 4.16b). The molecular dynamics model gave results that were consistent with the simpler kinematic for flat non-interacting substrates. Moving forward, this molecular dynamics setup will enable the investigation of substrates of industrially relevant materials and realistic surface roughness.

4.8 Mechanochemically Modified Networks

Mechanophore-based approaches are a promising yet challenging means for modifying polymer network architecture in response to force. Naturally, the first step to be considered in the design of any response is triggering the mechanophore, as discussed extensively in the preceding sections of the chapter. Briefly, mechanophore triggering is polymer and linking architecture dependent. In glassy polymers, mechanochemical reactions only take place during plastic deformation. In elastomers, large strains may be needed to trigger a response, but tailored architectures can be used to concentrate force within a particular molecular portion.[16,34,38] For example, in a polymerization or cross-linking reaction, the proximity of the reacting components to the activated mechanophore is critical. The polymer matrix must have sufficient mobility in order for a reaction to progress or the reacting components will rapidly deplete or never even initiate. To date, two mechanophore-modified networks and a few related steps have been published. We discuss these works below, including a perspective on the future of mechanochemically modified networks with a focus on the key themes of force and mobility.

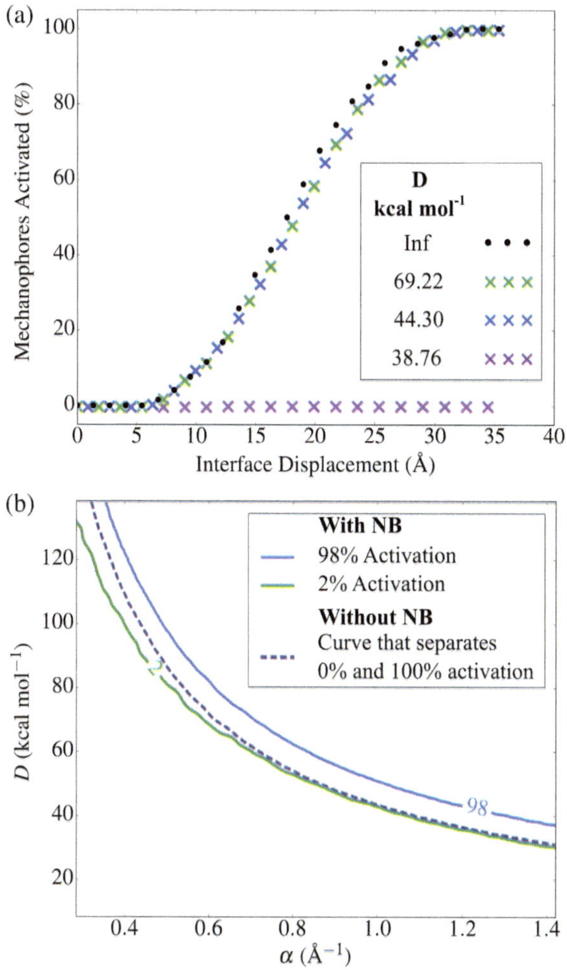

Figure 4.16 Kinematic model results. (a) Activation trends of mechanophores with different attachment potential depths *D*. Here "Inf" refers to a rigid attachment. (b) In the presence of non-bonded (NB) interactions the grid search performed of potential depth *D* and inverse well width *α* with the kinematic model shows a narrow band over which the transition from no activation to complete activation happens. Reprinted from *Extreme Mechanics Letters*, 8, M. S. Manivannan and M. N. Silberstein, Computational investigation of shear driven mechanophore activation at interfaces, 6–12, Copyright 2015,[66] with permission from Elsevier.

Jakobs *et al.* created an MCR polymer that polymerizes a monomer when a latent catalyst is activated by mechanical stress.[67] Specifically, the mechanophore is a latent metathesis catalyst with two tightly bound N-heterocyclic carbene (NHC) ligands on a ruthenium-alkylidene species previously developed within the Sijbesma group.[68] The mechanophore was covalently bonded within a high molecular weight semicrystalline poly(tetrahydrofuran) (pTHF) matrix. The pure pTHF polymer showed a leathery (between rubbery and glassy) stress–strain behavior under uniaxial compression (see Figure 4.17a) with an elastic modulus of 28 MPa and yield stress of 6 MPa. The semicrystalline matrix is necessary for transmission of sufficient force to unveil the catalyst. In order for a polymerization to occur once the catalyst is revealed, a compatible monomer needs to be present. A ratio of 15% mechanophore and 11% norbornene monomer by weight were added to the polymer (leaving 74% by weight pTHF). These additional components reduced the polymer

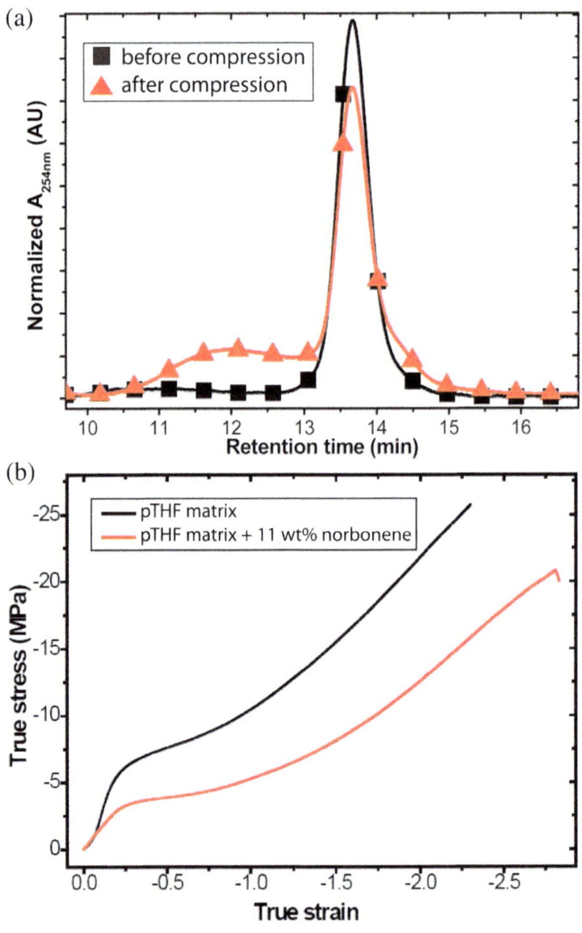

Figure 4.17 Results from pTHF-mechanocatalyst system designed to polymerize in response to mechanical force. (a) Stress–strain response under uniaxial compression (no mechanophore). (b) GPC trace showing growth of high-molecular-weight polymer following five cycles of compression in a pellet press to 800 MPa. Adapted with permission from R. T. M. Jakobs, S. Ma and R. P. Sijbesma, *ACS Macro. Lett.*, 2013, **2**, 613–616.[67] Copyright (2013) American Chemical Society.

mechanical stiffness and strength to 16 MPa and 3 MPa, respectively. Prior to polymerization, the monomer solvates the polymer and acts as a plasticizer, enhancing chain mobility. The effectiveness of the polymer-mechanophore system was proven with a pellet press based compression test. A spherical specimen was compressed under 0.8 GPa, folded and compressed four additional times to the same pressure. It is clear from the dimensions of the specimen after the pressure was removed that there is significant plastic deformation (plastic engineering strain around −0.9 after first compression). GPC traces showed the growth of a high-molecular-weight component of the polymer (see Figure 4.17b). The degree of conversion of the monomer to polymer depended on the applied pressure, number of cycles, to a small extent on the pressure hold time and to a significant extent on the total time from when the pressure is first applied. These results suggest that the catalyst activation happens quickly (seconds or less), but that the polymerization happens slowly (order of minutes). The degree of polymerization did not continue to evolve with longer times even though the catalyst should have remained active and there was still unreacted monomer. The authors suggested that this is the result of localized monomer depletion in the area around the catalyst. This local depletion is likely, particularly since mobility will decrease as the monomers are consumed and the material becomes more glassy. Utilizing a bifunctional monomer, the authors further demonstrated the ability to form cross-links in response to similar compression loading. Cross-linking is likely to be a more effective means for self-healing applications than uncross-linked chain polymerization in bulk.

Ramirez *et al.* created an MCR polymer that cross-links itself in response to chain scission level forces.[69] Specifically, the polymer is gDBC embedded in a PB backbone. This polymer/mechanophore combination had been shown in previous work by the authors to be mechanochemically active and susceptible to nucleophilic substitution.[70] The polymer was subjected to a shear stress applied by a twin screw extruder while the polymer was mixed with the tetrabutylammonium salt of sebacic acid that served as the cross-linker. Prior to and after extrusion, the mechanical properties (stiffness and hardness) were assessed with nanoindentation. The polymer had an elastic modulus of 8 MPa prior to shearing. The addition of the salt reduced the elastic modulus down to 2 MPa. After extrusion, the polymer had an elastic modulus of 150 MPa, nearly 100 times the pre-extrusion modulus. After extrusion and a week of waiting time the polymer had a modulus of 280 MPa. Although the Craig group demonstration differs from the Sijbesma group demonstration in mechanophore, mechanophore function, polymer matrix, initial polymer properties, triggered reaction, means of applying load and initial and final mechanical properties, the results are in many ways analogous. Both systems require forces that cause extensive damage to the material (*i.e.* not your typical tensile test), an additional component is required that enables the chemical reaction but also reduces the material stiffness and strength, significant polymer network formation occurs and mechanophore triggering happens much faster than the subsequent network formation reaction.

The Boydston and Moore groups each demonstrated a distinct MCR polymer with a small molecule release function. Diesendruck *et al.* incorporated the mechanophore gem-dichlorocyclopropane into cross-linked PMA.[71] This polymer system released an acid when the polymer was subjected to compressive loading within a pellet press. Activation increased with increased applied pressure. Larsen and Boydston incorporated the mechanophore oxanorbornadiene into cross-linked PMA and into a segmented PU system.[72,73] Small organic molecules were released when either solid-state polymer is subjected to compressive loading within a pellet press. Activation increased with applied pressure. For the PU system activation also increased with repeat compressions on a folded polymer (the cross-linked PMA was too brittle after compression for folding). For all three of these polymers, the molecule release (mechanophore activation) does not require scission of the host polymer chain.

It is worth noting that all of these mechanophore systems with function beyond mechanochromism have been demonstrated under compression (except for the Craig group system, which was demonstrated under an even more aggressive combined shear and compression loading), whereas the vast majority of MCR polymers have been demonstrated under tension. This is certainly in part due to ease of specimen preparation for a pellet press experiment, but compression is also better able to transmit force to a polymer even after fracture. MCR polymer theory would be helpful in advancing materials with stress-responsive mechanical and/or chemical properties so that they are active without requiring loads that induce wide scale damage, for instance by prescribing an appropriate cyclic loading history.

An additional mechanochemically adaptive network concept that has not yet been achieved is that of polymer chain contour length extension in response to force. Since the force on a polymer chain at a given stretch is determined by the chain contour length, contour length change is a natural way to modify the mechanical properties of the polymer. Large amounts of energy can be dissipated from length extensions – a technique that is commonly employed in biological systems. One well-studied example at the macroscale is the mussel byssus thread that attaches mussels to rocks.[74,75] The distal region of the thread extends, giving an apparent yielding behavior when it is subjected to wave impact. The thread then slowly refolds when the load is removed. This unfolding/refolding mechanism enables the mussel to survive large wave impacts while staying attached. If the length change is reversible then the energy can be dissipated without degrading the material (see Figure 4.18a). The Craig group has demonstrated this contour length extension concept for single polymer chains subjected to single molecular force microscopy (see Figure 4.18b).[9,59] Although they have made bulk polymers with the same mechanophores, the effect of mechanophore length extension on the mechanical response of the polymer has not been shown, likely due to the difficulty of triggering enough contour length change to effect the bulk response without damaging the polymer.

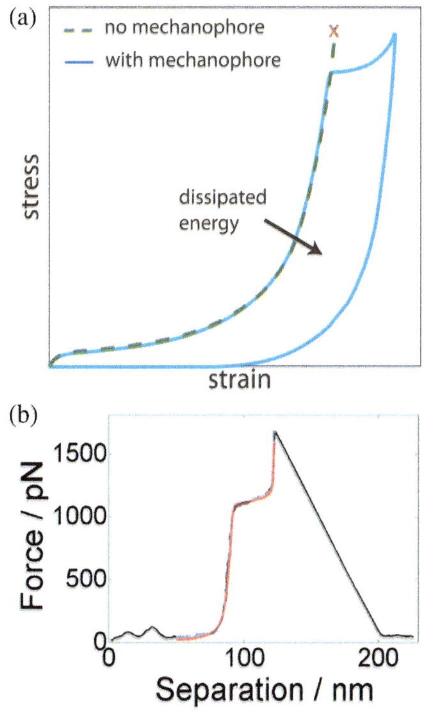

Figure 4.18 (a) Schematic of polymer network idea incorporating length-extending mechanophores to dissipate energy from contour length extension. (b) Length extension from mechanophore triggering demonstrated on a single polymer chain, reprinted with permission from D. Wu, J. M. Lenhardt, A. L. Black, B. B. Akhremitchev and S. L. Craig, *J. Am. Chem. Soc.*, 2010, **132**, 15936–15938.[59] Copyright (2010) American Chemical Society.

4.9 Conclusions

Extensive experimental and theoretical work has been performed on optically responsive MCR polymers. These optically responsive polymers have served as a valuable tool for understanding and quantifying how mechanophores trigger in polymers. The major themes that have arisen are stretch relative to the finite extensibility stretch, orientation, permanence of cross-links relative to the testing time scale and temperature relative to T_g. Researchers have, to some extent, been able to tailor when a mechanophore will activate; however, nearly always with changes to the polymer mechanical properties as well. One exception is the work of Chen *et al.* that used blends of mechanophore-linked polymers with plain polymers to tailor activation without significantly changing the overall mechanical properties.[35] Moving forward, heterogeneous polymers (blends, polyurethanes, copolymers, multi-networks) will be a natural medium for utilizing mechanophores since the internal structure intrinsically provides options for tailoring the onset of activation. These heterogeneous polymers should also make good test beds for new mechanophores. Composites have heterogeneity that can be used to enhance mechanochemical function, once the synthetic challenges are overcome. The interface between constituents in particular should induce activation at the early stages of damage.

The next frontier in MCR mechanochemistry is certainly moving beyond optical response to materials that have improved mechanical properties from mechanophore incorporation. While several initiatives have been taken toward this end, there is much to be done in terms of both theory and synthetic realization. Existing systems all require highly destructive forces. The theory for network adaptation should be able to draw heavily from existing work on gels,[76–78] dynamic covalent elastomers,[79–81] liquid crystal elastomers,[82] self-healing materials[83,84] and biological materials.[85,86] Experimental methods that enable *in-situ* monitoring of the mechanophore state and polymer mechanical properties for non-optical mechanophores will also accelerate development. Synthetically, options should expand as more mechanophores are developed and more research groups become familiar with their implementation.

Abbreviations

BDO	1,4-Butylene glycol
DABBF	Diarylbibenzofuranone
DN	Double network
DSC	Differential scanning calorimetry
EGDMA	Ethylene glycol dimethacrylate
EPR	Electron paramagnetic resonance
FTIR	Fourier transform infrared spectroscopy
GC-MS	Gas chromatography - mass spectroscopy
gDBC	Gem-dibromocyclopropane
GPC	Gel permeation chromatography
H NMR	Proton nuclear magnetic resonance
M_n	Number average molecular weight
M_w	Weight average molecular weight
NHC	N-heterocyclic carbene
NMR	Nuclear magnetic resonance spectroscopy
PA	Polyamide
PB	Polybutadiene
PCL	Polycaprolactone
PDMS	Polydimethylsiloxane
PE	Polyester
PEG550	Polyethylene glycol, 550 kDa molecular weight
PEG750	Polyethylene glycol, 750 kDa molecular weight
PMA	Poly(methyl acrylate)
PMMA	Poly(methyl methacrylate)
PnBA	Poly(*n*-butyl acrylate)

PS	Polystyrene
pTHF	Polytetrahydrofuran
PU	Polyurethane
SENT	Single edged notch test
SP	Spiropyran
TBA	Tetrabutylammonium salt of sebacic acid
T_g	Glass transition temperature
TN	Triple network
UPy	2-Ureido-4-pyrimidone
UV	Ultra violet
UV-Vis	Ultra violet - visible

Acknowledgements

This material is based in part upon work supported by the National Science Foundation (NSF) under grant no. DMR 1307354.

References

1. C. R. Hickenboth, J. S. Moore, S. R. White, N. R. Sottos, J. Baudry and S. R. Wilson, *Nature*, 2007, **446**, 423–427.
2. W. Kauzmann and H. Eyring, *J. Am. Chem. Soc.*, 1940, **62**, 3113–3125.
3. G. I. Bell, *Science*, 1978, **200**, 618–627.
4. P. Hanggi, P. Talkner and M. Borkovec, *Rev. Mod. Phys.*, 1990, **62**, 251–341.
5. D. A. Davis, A. Hamilton, J. L. Yang, L. D. Cremar, D. Van Gough, S. L. Potisek, M. T. Ong, P. V. Braun, T. J. Martinez, S. R. White, J. S. Moore and N. R. Sottos, *Nature*, 2009, **459**, 68–72.
6. V. I. Minkin, *Chem. Rev.*, 2004, **104**, 2751–2776.
7. G. O'Bryan, B. M. Wong and J. R. McElhanon, *ACS Appl. Mater. Interfaces*, 2010, **2**, 1594–1600.
8. G. R. Gossweiler, G. B. Hewage, G. Soriano, Q. M. Wang, G. W. Welshofer, X. H. Zhao and S. L. Craig, *ACS Macro Lett.*, 2014, **3**, 216–219.
9. G. R. Gossweiler, T. B. Kouznetsova and S. L. Craig, *J. Am. Chem. Soc.*, 2015, **137**, 6148–6151.
10. L. D. Cremar, *Insights for Designing Mechanochromic Spiropyrans from First Principles Dynamics and Minimum Energy Pathways*, PhD thesis, Stanford University, 2012.
11. Y. L. Chen, A. J. H. Spiering, S. Karthikeyan, G. W. M. Peters, E. W. Meijer and R. P. Sijbesma, *Nat. Chem.*, 2012, **4**, 559–562.
12. R. Gostl and R. P. Sijbesma, *Chem. Sci.*, 2016, **7**, 370–375.
13. Y. K. Song, K. H. Lee, W. S. Hong, S. Y. Cho, H. C. Yu and C. M. Chung, *J. Mater. Chem.*, 2012, **22**, 1380–1386.
14. K. Imato, T. Kanehara, T. Ohishi, M. Nishihara, H. Yajima, M. Ito, A. Takahara and H. Otsuka, *ACS Macro Lett.*, 2015, **4**, 1307–1311.
15. Y. G. Han, A. D. Rogalsky, B. X. Zhao and H. J. Kwon, *Polym. Eng. Sci.*, 2012, **52**, 826–834.

16. C. K. Lee, B. A. Beiermann, M. N. Silberstein, J. Wang, J. S. Moore, N. R. Sottos and P. V. Braun, *Macromolecules*, 2013, **46**, 3746–3752.
17. B. A. Beiermann, S. L. B. Kramer, P. A. May, J. S. Moore, S. R. White and N. R. Sottos, *Adv. Funct. Mater.*, 2014, **24**, 1529–1537.
18. L. R. G. Treloar, *Rep. Prog. Phys.*, 1973, **36**, 755–826.
19. W. Kuhn, *Kolloid-Z.*, 1936, **76**, 258–271.
20. M. C. Wang and E. Guth, *J. Chem. Phys.*, 1952, **20**, 1144–1157.
21. H. M. Klukovich, T. B. Kouznetsova, Z. S. Kean, J. M. Lenhardt and S. L. Craig, *Nat. Chem.*, 2013, **5**, 110–114.
22. J. Wang, T. B. Kouznetsova, Z. Niu, M. T. Ong, H. M. Klukovich, A. L. Rheingold, T. J. Martinez and S. L. Craig, *Nat. Chem.*, 2015.
23. H. Zhang, Y. J. Chen, Y. J. Lin, X. L. Fang, Y. Z. Xu, Y. H. Ruan and W. G. Weng, *Macromolecules*, 2014, **47**, 6783–6790.
24. C. M. Kingsbury, P. A. May, D. A. Davis, S. R. White, J. S. Moore and N. R. Sottos, *J. Mater. Chem.*, 2011, **21**, 8381–8388.
25. B. A. Beiermann, S. L. B. Kramer, J. S. Moore, S. R. White and N. R. Sottos, *ACS Macro Lett.*, 2012, **1**, 163–166.
26. G. I. Peterson, M. B. Larsen, M. A. Ganter, D. W. Storti and A. J. Boydston, *ACS Appl. Mater. Interfaces*, 2015, **7**, 577–583.
27. G. I. Peterson, M. Yurtoglu, M. B. Larsen, S. L. Craig, M. A. Ganter, D. W. Storti and A. J. Boydston, *Rapid Prototyping J.*, 2015, **21**, 520–527.
28. A. L. B. Ramirez, A. K. Schmitt, M. K. Mahanthappa and S. L. Craig, *Faraday Discuss.*, 2014, **170**, 337–344.
29. Q. M. Wang, G. R. Gossweiler, S. L. Craig and X. H. Zhao, *Nat. Commun.*, 2014, **5**, 4899.
30. Q. M. Wang, G. R. Gossweiler, S. L. Craig and X. H. Zhao, *J. Mech. Phys. Solids*, 2015, **82**, 320–344.
31. M. C. Boyce and E. M. Arruda, *Rubber Chem. Technol.*, 2000, **73**, 504–523.
32. J. M. Lenhardt, A. L. Black, B. A. Beiermann, B. D. Steinberg, F. Rahman, T. Samborski, J. Elsakr, J. S. Moore, N. R. Sottos and S. L. Craig, *J. Mater. Chem.*, 2011, **21**, 8454–8459.
33. C. K. Lee, D. A. Davis, S. R. White, J. S. Moore, N. R. Sottos and P. V. Braun, *J. Am. Chem. Soc.*, 2010, **132**, 16107–16111.
34. E. Ducrot, Y. L. Chen, M. Bulters, R. P. Sijbesma and C. Creton, *Science*, 2014, **344**, 186–189.
35. Y. L. Chen and R. P. Sijbesma, *Macromolecules*, 2014, **47**, 3797–3805.
36. X. L. Fang, H. Zhang, Y. J. Chen, Y. J. Lin, Y. Z. Xu and W. G. Weng, *Macromolecules*, 2013, **46**, 6566–6574.
37. Y. J. Chen, H. Zhang, X. L. Fang, Y. J. Lin, Y. Z. Xu and W. G. Weng, *ACS Macro Lett.*, 2014, **3**, 141–145.
38. S. C. Jiang, L. X. Zhang, T. W. Xie, Y. J. Lin, H. Zhang, Y. Z. Xu, W. G. Weng and L. Z. Dai, *ACS Macro Lett.*, 2013, **2**, 705–709.
39. M. N. Silberstein, L. D. Cremar, B. A. Beiermann, S. B. Kramer, T. J. Martinez, S. R. White and N. R. Sottos, *J. Mech. Phys. Solids*, 2014, **63**, 141–153.
40. J. L. Barrat, J. Baschnagel and A. Lyulin, *Soft Matter*, 2010, **6**, 3430–3446.
41. J. Rottler, *J. Phys.: Condens. Matter*, 2009, **21**, 463101.
42. O. A. Hasan and M. C. Boyce, *Polymer*, 1993, **34**, 5085–5092.

43. E. M. Arruda, M. C. Boyce and R. Jayachandran, *Mech. Mater.*, 1995, **19**, 193–212.
44. V. Srivastava, S. A. Chester, N. M. Ames and L. Anand, *Int. J. Plast.*, 2010, **26**, 1138–1182.
45. B. A. Beiermann, D. A. Davis, S. L. B. Kramer, J. S. Moore, N. R. Sottos and S. R. White, *J. Mater. Chem.*, 2011, **21**, 8443–8447.
46. C. M. Degen, P. A. May, J. S. Moore, S. R. White and N. R. Sottos, *Macromolecules*, 2013, **46**, 8917–8921.
47. J. W. Kim, Y. Jung, G. W. Coates and M. N. Silberstein, *Macromolecules*, 2015, **48**, 1335–1342.
48. C. K. Lee, C. E. Diesendruck, E. J. Lu, A. N. Pickett, P. A. May, J. S. Moore and P. V. Braun, *Macromolecules*, 2014, **47**, 2690–2694.
49. J. R. Hemmer, P. D. Smith, M. van Horn, S. Alnemrat, B. P. Mason, J. R. de Alaniz, S. Osswald and J. P. Hooper, *J. Polym. Sci., Part B: Polym. Phys.*, 2014, **52**, 1347–1356.
50. A. D. N. Celestine, B. A. Beiermann, P. A. May, J. S. Moore, N. R. Sottos and S. R. White, *Polymer*, 2014, **55**, 4164–4171.
51. M. E. Grady, B. A. Beiermann, J. S. Moore and N. R. Sottos, *ACS Appl. Mater. Interfaces*, 2014, **6**, 5350–5355.
52. K. Yoshimoto, T. S. Jain, K. V. Workum, P. F. Nealey and J. J. de Pablo, *Phys. Rev. Lett.*, 2004, **93**, 175501.
53. F. M. Capaldi, M. C. Boyce and G. C. Rutledge, *Polymer*, 2004, **45**, 1391–1399.
54. D. Hossain, M. A. Tschopp, D. K. Ward, J. L. Bouvard, P. Wang and M. F. Horstemeyer, *Polymer*, 2010, **51**, 6071–6083.
55. M. N. Silberstein, K. M. Min, L. D. Cremar, C. M. Degen, T. J. Martinez, N. R. Aluru, S. R. White and N. R. Sottos, *J. Appl. Phys.*, 2013, **114**.
56. M. T. Ong, J. Leiding, H. L. Tao, A. M. Virshup and T. J. Martinez, *J. Am. Chem. Soc.*, 2009, **131**, 6377–6379.
57. B. Isralewitz, M. Gao and K. Schulten, *Curr. Opin. Struct. Biol.*, 2001, **11**, 224–230.
58. J. Ribas-Arino and D. Marx, *Chem. Rev.*, 2012, **112**, 5412–5487.
59. D. Wu, J. M. Lenhardt, A. L. Black, B. B. Akhremitchev and S. L. Craig, *J. Am. Chem. Soc.*, 2010, **132**, 15936–15938.
60. J. K. Kim and Y. W. Mai, *Compos. Sci. Technol.*, 1991, **41**, 333–378.
61. J. Z. Liang and R. K. Y. Li, *J. Appl. Polym. Sci.*, 2000, **77**, 409–417.
62. N. Domun, H. Hadavinia, T. Zhang, T. Sainsbury, G. H. Liaghat and S. Vahid, *Nanoscale*, 2015, **7**, 10294–10329.
63. L. D. Vincent, *Protective Coating User's Handbook*, NACE International, Houston, Texas, 2nd edn, 2010.
64. K. Makyla, C. Muller, S. Lorcher, T. Winkler, M. G. Nussbaumer, M. Eder and N. Bruns, *Adv. Mater.*, 2013, **25**, 2701–2706.
65. S. Lorcher, T. Winkler, K. Makyla, C. Ouellet-Plamondon, I. Burgert and N. Bruns, *J. Mater. Chem. A*, 2014, **2**, 6231–6237.
66. M. S. Manivannan and M. N. Silberstein, *Extreme Mech. Lett.*, 2016, 6–12.
67. R. T. M. Jakobs, S. Ma and R. P. Sijbesma, *ACS Macro Lett.*, 2013, **2**, 613–616.

68. A. Piermattei, S. Karthikeyan and R. P. Sijbesma, *Nat. Chem.*, 2009, **1**, 133–137.
69. A. L. B. Ramirez, Z. S. Kean, J. A. Orlicki, M. Champhekar, S. M. Elsakr, W. E. Krause and S. L. Craig, *Nat. Chem.*, 2013, **5**, 757–761.
70. A. L. Black, J. A. Orlicki and S. L. Craig, *J. Mater. Chem.*, 2011, **21**, 8460–8465.
71. C. E. Diesendruck, B. D. Steinberg, N. Sugai, M. N. Silberstein, N. R. Sottos, S. R. White, P. V. Braun and J. S. Moore, *J. Am. Chem. Soc.*, 2012, **134**, 12446–12449.
72. M. B. Larsen and A. J. Boydston, *J. Am. Chem. Soc.*, 2013, **135**, 8189–8192.
73. M. B. Larsen and A. J. Boydston, *J. Am. Chem. Soc.*, 2014, **136**, 1276–1279.
74. K. Bertoldi and M. C. Boyce, *J. Mater. Sci.*, 2007, **42**, 8943–8956.
75. Z. Qin and M. J. Buehler, *Nat. Commun.*, 2013, **4**, 2187.
76. S. K. Ahn, R. M. Kasi, S. C. Kim, N. Sharma and Y. X. Zhou, *Soft Matter*, 2008, **4**, 1151–1157.
77. S. A. Chester, *Soft Matter*, 2012, **8**, 8223–8233.
78. Z. S. Liu, W. Toh and T. Y. Ng, *Int. J. Appl. Mech.*, 2015, **7**, 1530001.
79. R. Long, H. J. Qi and M. L. Dunn, *J. Mech. Phys. Solids*, 2013, **61**, 2212–2239.
80. R. Long, H. J. Qia and M. L. Dunn, *Soft Matter*, 2013, **9**, 4083–4096.
81. J. Ma, X. M. Mu, C. N. Bowman, Y. Y. Sun, M. L. Dunn, H. J. Qi and D. N. Fang, *J. Mech. Phys. Solids*, 2014, **70**, 84–103.
82. T. J. White and D. J. Broer, *Nat. Mater.*, 2015, **14**, 1087–1098.
83. J. Mergheim and P. Steinmann, *Comput. Mech.*, 2013, **52**, 681–692.
84. W. Binder, *Self-healing Polymers: From Principles to Applications*, 2013, pp. 1–3.
85. K. Garikipati, E. M. Arruda, K. Grosh, H. Narayanan and S. Calve, *J. Mech. Phys. Solids*, 2004, **52**, 1595–1625.
86. E. Kuhl, *J. Mech. Behav. Biomed. Mater.*, 2014, **29**, 529–543.

CHAPTER 5

Tailoring Mechanochemical Reactivity of Covalent Bonds in Polymers by Non-covalent Interactions

HUAN ZHANG, LINXING ZHANG, YINJUN CHEN, YANGJU LIN AND WENGUI WENG*

Department of Chemistry, College of Chemistry and Chemical Engineering, Xiamen University, Xiamen, Fujian 361005, P.R. China
*E-mail: wgweng@xmu.edu.cn

5.1 Introduction

Because of the recent development of polymer mechanochemistry, tuning the mechanochemical reactivity of mechanophores and the response of stress-responsive materials by non-covalent interactions poses a great challenge. We envision fruitful outcomes of these efforts, like new principles and novel designs, if the effect of non-covalent interactions on polymer mechanochemistry can be better understood. Herein we summarize recent studies in the use of supramolecular interactions to tune the mechanochemical reactivity of covalent bonds. We particularly focus on the achievements, limitations and prospects of this field. More general features of polymer mechanochemistry and stress-responsive materials can be found in other reviews.[1–8]

Polymer Chemistry Series No. 26
Mechanochemistry in Materials
Edited by Yoan C. Simon and Stephen L. Craig

Published by the Royal Society of Chemistry, www.rsc.org

5.2 Theoretical Background of Polymer Mechanochemistry and the Effects of Supramolecular Interactions

5.2.1 Coil-to-stretch Transition and Bead-rod Model

Theoretical studies of polymer mechanochemistry have been an active area since the late 1970s. De Gennes first predicted that a polymer coil would be drastically elongated to a fully extended state (coil-to-stretch transition) when the stretching exerted by the flow field conquered the relaxation of the chain (Figure 5.1).[9] Odell and Keller pioneered a bead-rod model to explain the mechanochemical cleavage of an isolated chain in dilute solution.[10,11] This model is the basis for further discussion.

In the bead-rod model, a chain with $N = 2m + 1$ beads is in fully stretched conformation, *i.e.* a rod. Each bead experiences a hydrodynamic drag force from the surrounding medium. To compensate for the hydrodynamic drag force, internal tension accumulates from the chain end to the middle point. In the bead-rod model, the tension force on the ith bead σ_i has a parabolic distribution along the chain contour (see Figure 5.1 for coordinate):

$$\sigma_i = \sigma_{\max}\left(\frac{m^2 - i^2}{m^2}\right) \tag{5.1}$$

$\sigma_{\max}$ is the maximum tensile force at the middle point ($i = 0$) and has an expression:

$$\sigma_{\max} = \frac{3}{4} S_1 \pi \dot{\varepsilon} a b \eta_s N^2 \tag{5.2}$$

where a is the radius of the beads, b is the length of the chain segment, η_s is the solvent viscosity, S_1 is the "shielding factor", which accounts for

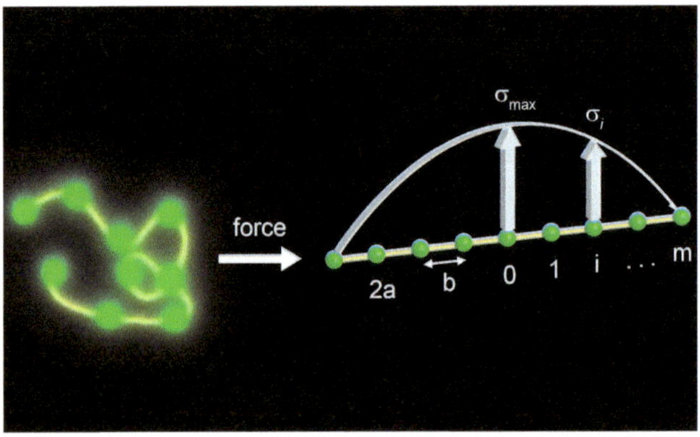

Figure 5.1 Schematic illustration of the coil-to-stretch transition of a polymer chain in the flow field. The internal tension σ_i has a parabolic distribution along the chain contour (bead-rod model).

hydrodynamic interactions between beads and is often set to 1 in the *fully stretched* conformation.[12]

Chain scission occurs when σ_{max} exceeds a critical value f_{sci}. In fact, f_{sci} calculated from eqn (5.1) is well in line with the bond strength obtained from quantum chemical calculation.[10,11,13] For given polymers (a and b) and experimental conditions (η_s and strain rate), σ_{max} scales with the square of N (or the molecular weight M). Therefore, shorter polymers with $M < M_{lim}$ cannot be mechanically fractured compared to their longer counterparts. The same rule is valid for the activation of mechanophores embedded in the polymer backbone.[14,15]

5.2.2 Intra-chain Non-covalent Interactions

Polymer mechanochemistry is a multi-scale phenomenon, ranging from macroscopic deformation to the extension of individual polymer chains to the fraction of covalent bonds. Therefore, the effects of non-covalent interactions on polymer mechanochemistry also can be considered at different length scales.

We start our discussion with the mechanochemistry of an isolated polymer chain bearing multiple supramolecular motifs (Figure 5.2a). This constraint implicitly limits our discussion to *dilute solutions*. The relevant outcomes of

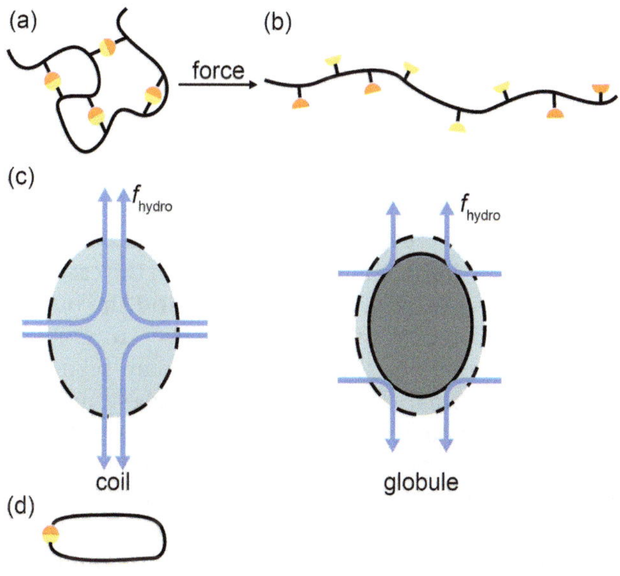

Figure 5.2 Effects of intra-chain supramolecular interactions on the mechanochemical reaction of an isolated polymer chain. (a) A chain has multiple supramolecular motifs. (b) The corresponding chain extension breaks non-covalent bonds. (c) Both the inner and outer parts of a polymer coil experience hydrodynamic drag force, f_{hydro}, whereas only the outer periphery of a globule feels f_{hydro}. The inner parts are in flow stagnation because of supramolecular interactions. (d) The supramolecular interaction leads to a cyclic conformation in the fully extended state.

the bead-rod model are the coil-to-stretch transition and sufficiently large σ_{max} to overcome f_{sci}. As a result, the roles of supramolecular interactions at this length scale can be understood from the following aspects.

5.2.2.1 *Sacrificial Units to Delay Chain Extension*

Because of the associations between the supramolecular motifs, the isolated polymer chain exhibits more compact globular shapes than the random coil conformation. The corresponding chain extension in the flow field is referred to as the "globule-stretch" transition.[16–18] The coil-to-stretch transition simultaneously occurs as the hydrodynamic drag force overcomes the entropic force, whereas the globule–stretch transition requires additional hydrodynamic drag force to break the supramolecular motifs (Figure 5.2b). Alexander-Katz and coworkers investigated the globule–stretch transition by computer simulation.[16–19] They found the critical strain rate to initiate chain extension increased as a function of the binding strength between supramolecular motifs in both shear and elongational flow fields. For example, the normalized critical strain rate changed from 0.001 to 0.08 as the normalized interaction energy increased from 0.41 to 2.09 for 50-bead chains in the case of free-draining (Figure 1 of ref. 18), respectively.[18]

5.2.2.2 *Additional Hydrodynamic Shielding to Delay Chain Extension*

Another way for supramolecular interactions to delay chain extension is to provide additional hydrodynamic shielding. The hydrodynamic drag force stems from the friction of the moving chain segment with respect to the surrounding solvent molecules. The moving chain segment in turn disturbs the flow profiles. These perturbations propagate to nearby chain segments, and are referred to as hydrodynamic interactions. In the case of a coil, hydrodynamic shielding is weak, and both the inner and the outer parts of the coil experience hydrodynamic drag force. However, in globular polymers, the hydrodynamic interactions shield the inner chain segments because of the more compact conformation. In Alexander-Katz's simulation,[18] the globule behaves like a hard sphere, and only the outer periphery of the globule feels the hydrodynamic drag force (Figure 5.2c). Only by allowing protrusions of chain segments to extend beyond the shielded area did the fluid exert enough force to stretch the molecule to its full contour length. This shielding effect was more profound at higher values of binding strength.

5.2.2.3 *Reducing σ_{max} at the Fully Extended State*

Supramolecular motifs are able to alter the chain topology, *e.g.* the ring-chain equilibrium.[20] If the loop conformation can be retained in the fully stretched state, the contour length exposed to the flow field (N in eqn (5.1)) is reduced. For the sake of convenience, we consider an isolated chain with supramolecular motifs at the two ends to form a ring (Figure 5.2d). In the extended conformation, such ring chain is equivalent to a linear chain having 1/2 N beads. The corresponding σ_{max} is only one-quarter as large as the original chain (eqn (5.1)). Cifre and coworkers studied the coil-to-stretch transition of cyclic polymer solution using Brownian dynamic simulation techniques.[21] The critical strain rate to induce coil-to-stretch transition for the ring chain was about four times that for the linear counterpart. The simulation is quantitatively consistent with the bead-rod prediction. The authors thus concluded that the ring chain was more resistant to mechanical degradation. This conclusion may extend to polymers having multiple binding sites once the loop conformations are partially retained in the stretched state.

5.2.3 Inter-chain Non-covalent Interactions

5.2.3.1 Linear Supramolecular Polymers

Next we consider polymer aggregates formed by inter-chain supramolecular interactions. Figure 5.3a shows a linear supramolecular polymer chain that is composed of oligomers or prepolymers held together by non-covalent bonds. This type of polymer chain can respond reversibly to external stimuli and has received great attention in the last decade.[22–26] Linear supramolecular polymers have an effectively higher molecular weight than the constituent chains, and thus are believed to have greater mechanochemical reactivity.

Imagine a chain ($N = 2m + 1$ beads) that associates at both ends with two additional chains ($N' = 2n$ beads) to form a linear supramolecular chain ($N + 2N'$ beads, Figure 5.3b). σ_{max} increases to:

$$\sigma_{max} = \frac{3}{4} S_1 \pi \dot{\varepsilon} a b \eta_s (N + 2N')^2 \quad (5.3)$$

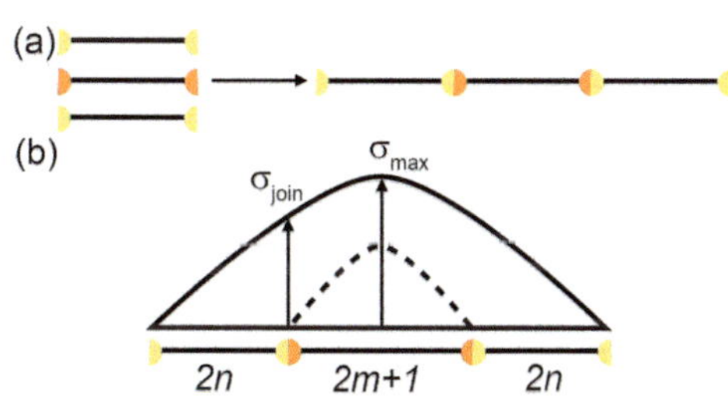

Figure 5.3 Mechanochemistry of linear supramolecular polymers. (a) End-capped oligomers or prepolymers form a linear supramolecular polymer. (b) The profile of the internal tension for the linear supramolecular chain. The dashed line represents the tension profile of the middle chain before association with the two side chains.

However, this expression is valid only if the chain does not dissociate at the binding sites. According to eqn (5.1), the tension force at the joints between the individual oligomers, σ_{join}, is expressed by:

$$\frac{\sigma_{join}}{\sigma_{max}} = 1 - \frac{1}{(1 + 2n/m)^2} \tag{5.4}$$

If all constituent chains have the same length ($n = m$), $\sigma_{join}/\sigma_{max} = 8/9$. This requires that the non-covalent bonds at the joint points are almost as strong as the covalent bonds. This is not feasible, because non-covalent interactions are usually weaker than covalent bonds at the force-free state, let alone when under mechanical force. In the case of mechanophores, the force required for activation is much lower. Therefore, it may be possible to use non-covalent interactions to enhance the activation of mechanophores through the formation of linear supramolecular polymer. From eqn (5.4), short side chains (n), long middle chain (m) and strong binding motifs can reduce the ratio $\sigma_{join}/\sigma_{max}$ while increasing σ_{max}. In short, supramolecular polymers have mechanical stability inferior to that of conventional polymers of the same chain length.

5.2.3.2 Ladder-like Polymer Aggregates

If the constituent chains have multiple supramolecular motifs, the formation of topological complex aggregates as a consequence of inter-chain interactions is favored. Agarwal and Mashelkar considered a model system where two fully extended chains (each have N beads) held by secondary interactions arrange parallel to each other along the flow direction (Figure 5.4).[27] The supramolecular motifs tied the two chains together but did not act as load-bearing units because they were perpendicular to the flow direction. The authors found that the σ_{max} of the ladder polymer was smaller than the linear counterpart having the same number of beads ($2N$) because of the change of topology (as shown in Section 5.2.2.3). Most importantly, the neighboring chain provided additional hydrodynamic shielding of hydrodynamic drag force. Accounting for both effects, σ_{max} of each component chain was smaller in the aggregates than in isolated conformation. Numerical calculation reported up to a 35% increase in the limiting molecular weight M_{lim}. The gain could be higher if more polymer chains are arranged to form the ladder-like aggregate.

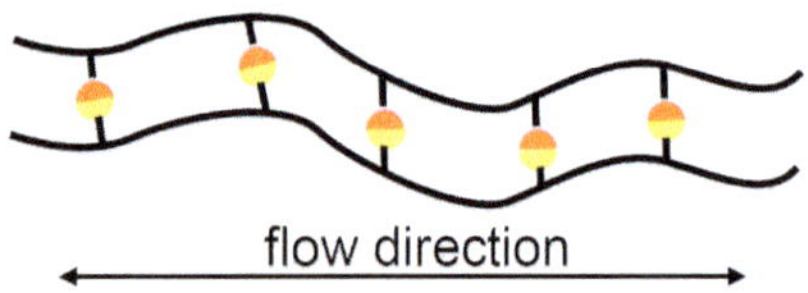

Figure 5.4 Two chains that have formed a ladder-like aggregate through supramolecular binding along the flow direction.

The key points in Agarwal's model are the redistribution of the hydrodynamic drag force and the additional shielding from a neighboring chain because of supramolecular interactions. These statements are similar to what we have proposed in Section 5.2.2. However, these conclusions still require more modelling, simulation and experimental evidence.

5.2.4 Non-covalent Interactions in the Bulk

In the bulk, inter-chain supramolecular interactions lead to the formation of long chain supramolecular polymers or cross-linked networks, which drastically reduces the chain dynamics and favors force transmission. The associated supramolecular motifs may have different polarity from the rest of the polymer backbone. In addition, they may form crystalline or pseudo-crystalline areas and/or phase-separate from other regions. Namely, non-covalent interactions can alter the condensed state of polymer chains, *i.e. morphological effects*. Examples include supramolecular units incorporated into block copolymer and polyurethane thermoplastics.[15,23,28–30] The supramolecular network, the crystalline or pseudo-crystalline and the micro-phase-separated structure completely alter the load-bearing properties and in turn the mechanochemistry of the materials in the bulk, as shown in the following sections.

5.3 Experimental Observations

5.3.1 Mechanical Cleavage of Covalent Bonds in Dilute Solution

5.3.1.1 Shear Stability of Supramolecular Polymers and Aggregates in Drag Reduction

Scientists have long known that a small amount of polymers (less than 100 ppm) can drastically reduce the drag in turbulent flow compared to the values observed for pure solvent.[31–33] Polymers with longer chain lengths possess superior drag reduction capability than their shorter cousins. However, the probability of chain breakage also increases with increasing molecular weight, as shown by eqn (5.1). Their degradation under intense flow greatly limits their utility for reducing drag in flow applications. Because drag reduction has huge economic benefits in applications like oil transportation, flood water disposal and fire-fighting,[34] there have been significant investigations into the development of shear stable polymeric systems since the 1960s. Common strategies include altering the chain topology by grafting on side chains,[35–37] mixing polymers with surfactants or fibers,[38–41] and introducing supramolecular associations.[42–45]

The idea of using supramolecular motifs is motivated by three factors. (1) Polymers having supramolecular motifs are capable of forming intra- and/or inter-chain associations. The extension of either of them has to break the non-covalent bonds in preference to the cleavage of polymer backbone (see Sections 5.2.2 and 5.2.3). (2) Supramolecular aggregates have a higher molecular weight than the constituent chains, and thus possess better drag reduction capabilities. (3) Even if the aggregate breaks, then the non-covalent bonds can still reform again in weaker flow regions to regain the original drag reduction efficiency.

In one example, Ting and Little[35] studied the shear stability of polyacrylic acid (PAA) sodium salt. At a Reynolds number of 9000, a polyethylene oxide (PEO) control lost about 70% of its drag reduction ability after 8 passes because of chain degradation. In stark contrast, the drag reduction efficiency of two PAA solutions was almost constant. The authors argued that intra-chain associations between two adjacent COO^- groups and one Na^+ ion (Figure 5.5a) formed a "virtual chain" parallel to the polymer backbone. The hydrodynamic drag force broke the temporary bonds prior to the covalent bonds and the temporary bonds reformed in the lower shear regions, accounting for the improved shear stability.

Although PAA sodium salt showed better shear stability than PEO, its drag reduction efficiency was not as good as PEO, most likely because the

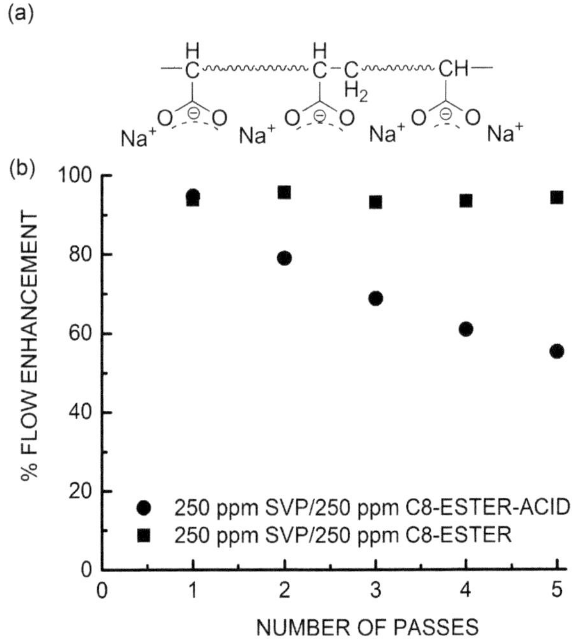

Figure 5.5 (a) Proposed ladder-like structure in ionized PAA. The shear stability of an associating SVP/C8-ESTER-ACID complex and its non-associating SVP/C8-ESTER analog. (b) Reprinted from *Journal of Non-Newtonian Fluid Mechanics*, 24(1), R. M. Kowalik, I. Duvdevani, D. G. Peiffer, R. D. Lundberg, K. Kitano and D. N. Shulz, Enhanced drag reduction *via* interpolymer associations, 1–10, Copyright (1987) with permission from Elsevier.[43]

intra-chain association led to more compact conformations.[27] To overcome these issues, other researches have turned to inter-chain associations. In the report from Kowalik and coworkers, the authors prepared three copolymers: (i) poly(octene-*co*-methyl undecenoate) (C8-ESTER); (ii) poly(octene-*co*-methyl undecenoate-*co*-undecenoic acid) (C8-ESTER-ACID); and (iii) poly(styrene-*co*-vinylpyridine) (SVP).[43] Inter-chain bindings were formed between SVP and anionic C8-ESTER-ACID through acid/base interactions, but were not formed in the SVP/C8-ESTER mixture. The drag reduction activity of the non-associative SVP/C8-ESTER dropped remarkably because of the chain scission (Figure 5.5b). In sharp contrast, SVP/C8-ESTER-ACID was much more shear stable in the flow field. The authors ascribed the result to the preferential cleavage of intermolecular associations to prevent backbone scission in the high shear region.

Mashelkar and coworkers tested the degradation of polymer backbone in the supramolecular aggregates after drag reduction.[45] The authors synthesized a proton donating copolymer (PD), poly(dodecyl acrylate-*co*-methacrylic acid), and a proton accepting copolymer (PA), poly(dodecyl acrylate-*co*-styrene-*co*-vinyl pyridine). When PD was mixed with PA, a compact supramolecular network (similar to the illustration shown in Figure 5.4) was formed because of the hydrogen bonding interactions between methacrylic acid and vinyl pyridine. At a capillary flow rate of 160 000 s^{-1}, the inter-chain complex showed improved shear stability over the nonassociative control. After the flow was stopped, the sheared sample was precipitated, dried and redissolved in toluene. The relative viscosity of the unsheared samples and the sheared samples were 0.76 *vs.* 0.75, respectively. The identical relative viscosity suggests no cleavage of backbone of the inter-chain complex. Unlike previous researchers, Mashelkar and coworkers believed supramolecular motifs did not act merely as sacrificial units in drag reduction (see Section 5.2.2.2).[27]

5.3.1.2 *Effect of Supramolecular Interactions on the Mechanochemical Reactivity of Disulfide Bond in Biomacromolecules*

Biomacromolecules have folded domains held by supramolecular interactions. Mechanical force can drive the conformation changes of the folded domains and regulate the physiological functions.[46] Examples of this type of phenomenon include the folding and refolding of proteins,[47,48] tuning mechanical stabilities of tandem arrays,[49,50] upregulating enzyme activity[51] and changing fluorescence resonance energy transfer between fluorescent proteins.[52] In this section, we take the mechanochemical reactivity of the disulfide/thiol exchange in proteins as an example to explore the influence of supramolecular interactions.

Disulfide bonds exist in many proteins for maintaining their tertiary structure. Discher and coworkers studied the disulfide/thiol exchange during the forced unfolding of the vascular cell adhesion molecule-1

(VCAM-1) protein.[53] VCAM-1 has seven immunoglobulin (Ig) domains, and each domain has a disulfide bond buried in its core (Figure 5.6a). The protein was adsorbed onto mica and stretched by atom force microscopy (AFM). The authors performed three types of experiments. First, when the protein was extended without reducing agent, a sawtooth pattern was observed in the force–extension curve, where each peak-valley corresponded to the unfolding of an Ig domain. Second, when the protein was not subject to extension and the mica was washed with reducing agent (iodoacetaminde), no reduction of the disulfide bonds was observed. Finally, when the protein was extended in the presence of the reducing agent, the disulfide bonds were reduced to thiols. Their results imply that mechanical force is able to regulate the redox state of disulfide bonds. Fernandez and coworkers employed the same technique to study the bond elongation of disulfide bonds embedded in engineered polyproteins.[54] The reduction rate increased as a function of applied force in the range of 0 to 500 pN for various reducing agents. Fitting the data to the Arrhenius model yielded the distance between the transition state (TS) and the reagent, $\Delta x^{\ddagger}$, in the range of 0.23 to 0.46 Å. The authors interpreted $\Delta x^{\ddagger}$ as the elongation of the disulfide bond under force. Fernandez and coworkers recently increased the applied force to 2 nN.[55] However, they observed a less pronounced rate-force dependence of disulfide/thiol exchange when the force was above 500 pN.

The force-regulated kinetics of thiol/disulfide exchange may involve at least two mechanisms: (i) the folded Ig domains act like a "gate" (similar to those discussed in Sections 5.2.2.1 and 5.2.2.2), *i.e.* the disulfide bonds will not be attacked by reducing agent before the Ig domains are unfolded by force; (ii) the molecular distortion and elongation of the disulfide moiety enhances its reactivity. The latter is an intrinsic property of disulfide moiety. Boulatov and coworkers tested the second mechanism using a molecular force probe, stiff stilbene (Figure 5.6b).[56] The *E*/*Z* isomerization of stiff stilbene introduces molecular force to the disulfide moiety incorporated between the C6 and C6′ atoms. This molecular force is tunable by the length of the linkers, *X* and *Y*, and is calculable.[57] The authors revealed that the activation enthalpies of thiol/disulfide exchange differed by less than 1 kcal mol^{-1} between the strained (*E*) and strain-free (*Z*) isomers (Figure 5.6c). As a result, the thiol/disulfide reaction was almost insensitive to the applied force up to 350 pN.[56] Aided by quantum chemical calculation, Boulatov and coworkers demonstrated that both the elongation of the –S–S– bond and the contraction of the –C–S–S– and the –S–S–C– bond angles should be considered to understand its mechanochemical reactivity. Their result questions the contribution of the second mechanism, but supports the first mechanism in the acceleration seen in stretched proteins. Although still under debate, this example suggests a strategy to tune the mechanochemical reactivity of covalent bonds by using supramolecular motifs as a "gate".

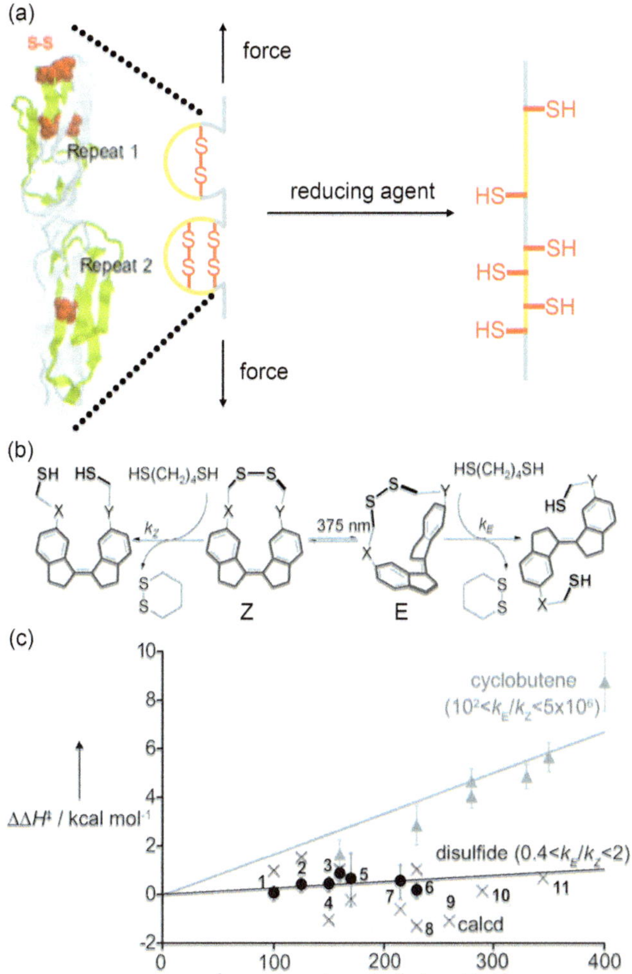

Figure 5.6 Mechanochemical reactivity of the disulfide/thiol exchange. (a) Force-induced reduction of disulfide bonds buried in the Ig domains of VCAM-1 protein. For clarity, only two Ig domains are displayed. This research was originally published in *Journal of Biological Chemistry*, N. Bhasin, P. Carl, S. Harper, G. Feng, H. Lu, D. W. Speicher and D. E. Discher, Chemistry on a single protein, vascular cell adhesion molecule-1, during forced unfolding, 2004, **279**(44), 45865–45874. Copyright (2004) the American Society for Biochemistry and Molecular Biology (ref. 53). (b) The use of a molecular force probe, stiff stilbene to explore the disulfide/thiol exchange. The molecular force exerted on the disulfide moiety can be tuned by the length of the X and Y spacers. (c) Difference in the activation enthalpies (300 K) between the Z and E isomers of the same macrocycle as a function of the calculated restoring force along the scissile S–S bond. ●: measured. ×: calculated by BELYP/6-311G**(CPCM-UA0). Reprinted with permission from T. J. Kucharski *et al.*, Kinetics of thiol/disulfide exchange correlate weakly with the restoring force in the disulfide moiety, *Angewandte Chemie*, John Wiley and Sons. Copyright © 2009 WILEY-VCH Verlag GmbH & Co. KGaA, Weinheim.

5.3.2 Effects of Strong Hydrogen Bonding Interactions on the Activation of Mechanophores in the Bulk

5.3.2.1 *Polymers Containing Both Supramolecular Motifs and Mechanophores on the Backbone*

The Weng group first incorporated both the mechanochromic covalent spiropyran (SP) mechanophore (Figure 5.7a) and the quadruple-hydrogen-bonding ureidopyrimidinone (UPy) motif[22] (Figure 5.7b) into the same polymer backbone (Figure 5.7c) *via* polyurethane (PU) polymerization.[58] The self-complementary UPy motifs have a very high binding

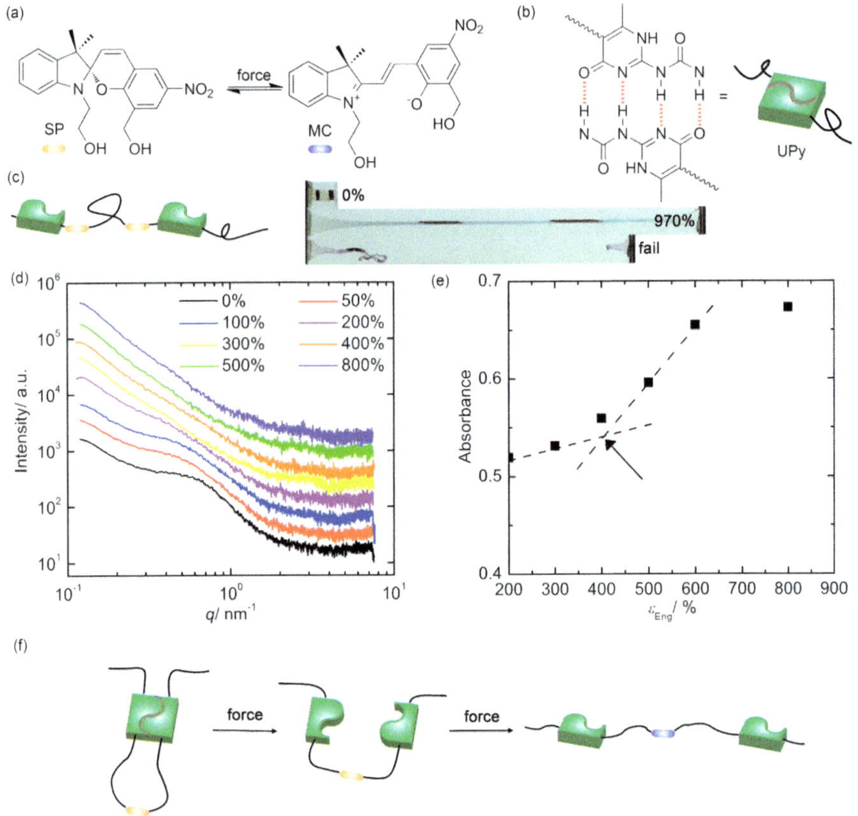

Figure 5.7 Effects of supramolecular interactions on the mechanical activation of linear supramolecular polymers. (a) Chemical structure of SP and MC. (b) Chemical structure of UPy. (c) Schematic drawing of the main-chain supramolecular polymer. (d) SAXS of the material under various strain indicated in the legend. (e) UV absorbance at 585 nm as a function of strain. The arrow indicates the onset of SP activation. (f) A possible loop formed as a consequence of intra-chain dimerization of UPy motifs. Reprinted with permission from ref. 58. Copyright 2013 American Chemical Society.

constant ($6 \times 10^7\ M^{-1}$ in $CDCl_3$[59]) to form dimers, and the dimers have been shown to further assemble into stacks with the aid of lateral hydrogen bonds from neighboring urea and urethane groups.[28] Because of thermodynamic incompatibility, the UPy stacks aggregate and phase-separate from the poly(tetrahydrofuran) (PTHF) matrix, resulting in soft/hard morphology.[60] As the sample stretched above a critical value strain (*ca.* 400%), the color of the specimen changed from yellow to pale blue because of the force-induced ring-opening reaction of SP to merocyanine (MC). The blue color intensified in the whole gauge region but changed to purple after catastrophic failure (Figure 5.7c). The two color changes suggest the presence of different isomers of MC (*vide infra*).

This polymer not only exhibits superb mechanical properties, but also offers a model to study the sequential responses of supramolecular motifs and covalent mechanophore under mechanical force. The UPy dimers, the UPy stacks (the hard domains) and the covalent SP moieties are all mechanically labile units. The authors started by asking the question: which mechanically triggered failure mechanism occurs first in response to external loading, the disruption of the UPy stacks, the dissociation of the UPy dimers or the activation of SP? At the strain-free state, the microphase-separation caused a primary scattering peak at $q = 0.6\ nm^{-1}$ because of inter-domain interference as revealed by small-angle X-ray scattering (SAXS, Figure 5.7d). However, this peak started to diminish at a strain of 200% and completely disappeared at 400%, suggesting the fracture of UPy stacks. On the other hand, MC has a strong UV absorbance at 585 nm, and this characteristic peak abruptly increased in intensity above a threshold strain of 400% (Figure 5.7e). Last, but most importantly, the reported force to break the UPy dimer is 100–200 pN,[61] slightly less than but comparable to the value reported for the activation of SP (240–260 pN[62]). Collectively, the hard domains, as well as the UPy stacks, fracture first, followed by the dissociation of UPy dimers, and, eventually, the activation of SP. It should be noted that the dimerization of UPy can inevitably lead to the formation of intra-chain loops in which the SP mechanophores are embedded (Figure 5.7f), therefore, the UPy dimers are expected to behave as "gate" for the SP mechanophores in loops.

5.3.2.2 *Mechanochemistry of SP in Soft and Hard Segments of PU*

In the design reported by Weng and coworkers, SP moieties might be incorporated in both the soft and hard segments of PU. What if SP is only incorporated into one segment? To answer this question, Braun and coworkers selectively incorporated SP moieties into either the soft segment (SPSS) or hard segment (SPHS) in PU.[63] They set the hard segment composition at two values: a low 22 wt% and a high 40 wt%. The mechanical properties of the materials were determined by the hard segment composition, regardless of the location of SP. The authors employed *in situ* fluorescence spectroscopy to monitor the force-induced activation of SP to MC. The normalized fluorescence I/I_0 decreased and then increased above a critical stretch ratio (Figure 5.8a). In the 22 wt% materials, SPSS and SPHS exhibited comparable I/I_0, whereas SP activation was always more predominant in SPSS than SPHS in the 40 wt% materials (Figure 5.8b). The authors further studied the orientation of SP moieties as a function of strain. They measured the total fluorescence intensity with the linear polarizer parallel, $I_{\parallel}$, and perpendicular to the stretch direction, $I_{\perp}$. The relative orientation of SP was defined using an order parameter $<P_2>$:

$$<P_2> = \frac{<3\cos^2(\beta) - 1>}{2} = \frac{I_{//} - I_{\perp}}{I_{//} + 2I_{\perp}} \tag{5.5}$$

where β is the angle between the dipole of MC and the macroscopic tensile axis. $<P_2>$ is 0 for random distribution and 1 for perfect alignment. At strain-free state, $<P_2>$ was nearly 0 for all the samples and then increased

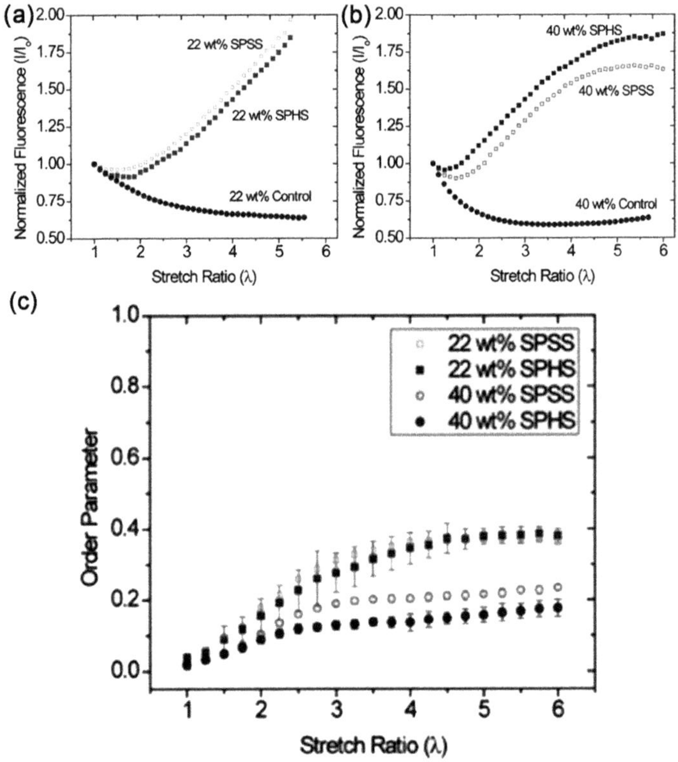

Figure 5.8 Activation of SP in the soft and hard segments of PUs. Normalized fluorescence during monotonic stretching of PU with (a) 22 wt% hard segment and (b) 40 wt% hard segment. (c) Mechanically activated order parameter *vs.* stretch ratio of SPSS and SPHS for both 22 and 40 wt% PU. Reprinted with permission from ref. 63. Copyright 2013 American Chemical Society.

monotonically as a function of stretch ratio. It should be noted: (i) for a given stretch ratio, $\langle P_2 \rangle$ for the 22 wt% materials was larger than that for the 40 wt% counterparts; (ii) SPSS and SPHS emerged with the same degree of orientation in the 22 wt% PU whereas SPSS was more oriented than SPHS (0~0.15 *vs.* 0~0.10) in the 40 wt% PUs. These results are in line with those shown in Figure 5.8a and b.

The authors ascribed the different levels of SP orientation to the different microstructures. In 22 wt% PU, the hard segments are partially miscible with the soft segments and the load is distributed uniformly. Therefore, the SP moieties in either the soft segment or hard segment have sufficient degree of freedom to rotate and align along the tensile direction. However, in the 40 wt% PU, phase-separation of the soft segment and hard segment was inferred based on DSC measurements. The hard domains are unlikely to be deformed affinely as the soft domains. Moreover, the high melting temperature (160 °C) of the hard domains constrains the local segment dynamics and thus hinders the orientation of SP. Therefore, SPHS showed lower value of $\langle P_2 \rangle$ than SPSS in 40 wt% materials and $\langle P_2 \rangle$ of 40 wt% PUs was smaller than that of 22 wt% PUs.

Braun and coworkers' study addresses an important aspect of supramolecular interactions in polymer mechanochemistry besides load-bearing unit. If a high level of mechanophore activation is desirable, it is important to ensure that the supramolecular interactions do not restrict the mobility of the chain segments with embedded mechanophores.

5.3.2.3 *Dioxetanes as Mechanoluminescent Probes in Thermoplastic Elastomers*

SP is a non-scissile type of mechanophore with a relatively low activation force compared to other mechanophores having three- or four-membered rings.[62,64] Thus, its activation does not correlate with the scission of polymer backbone. In contrast, 1,2-dioxetane is a scissile type of mechanophore first reported by Sijbesma and coworkers (see Figure 5.9a).[65] The strained four-membered ring is cleaved through a cycloreversion reaction into two carbonyl moieties, one of which is in an electronically excited state and can emit blue light upon relaxation. The cumulative light intensity therefore directly scales with the number of chain scission events and is strongly affected by the glass transition temperature of the polymer matrix.[65]

Recently, Sijbesma and coworkers studied the effect of hydrogen bonding on the mechanoluminescence of 1,2-dioxetane in segmented thermoplastic elastomers.[66] The authors incorporated bis(3-aminopropyl)-1,2-dioxetane in the hard segment of PU, polyester (PE) and polyamide (PA) of similar molecular weight (*ca.* 12 kDa, Figure 5.9b). Each elastomer also incorporated a soft segment, which consisted of poly(tetrahydrofuran) (PTHF). Because the PE film was extremely brittle, PU, PA and PE were mixed into a PU matrix (42 kDa) without embedded mechanophores (mPU)

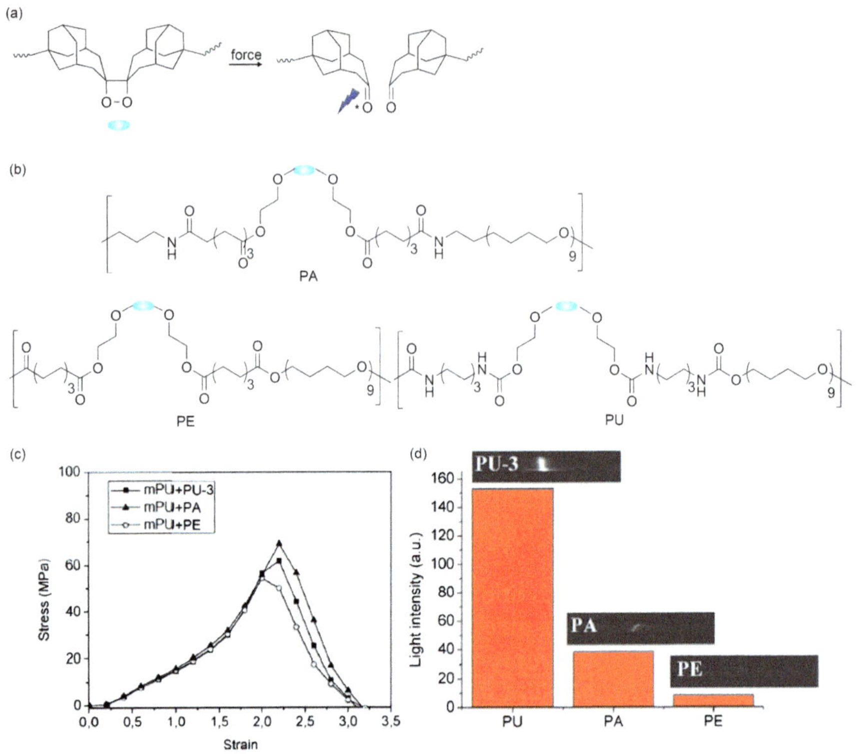

Figure 5.9 Effect of hydrogen bonding on the mechanical activation of 1,2-dioxetane in PU, PA and PE embedded in plain PU (mPU). (a) Chemical structure of 1,2-dioxetane and its mechanoluminescence. (b) Chemical structure of PU, PA and PE. (c) Stress–strain curves of mixed films doped with 0.5 wt% 4.7-di(thiophen-2-yl)benzo[*c*][1,2,5]thiadiazole. (d) The cumulative light intensity of the films until failure. The inlets are the corresponding optical images. Reprinted with permission from ref. 66. Copyright 2014 American Chemical Society.

to fabricate elastic films. All mixed films showed similar stress–strain curves at a strain rate of 20 s^{-1} (Figure 5.9c), but differed in light intensity (Figure 5.9d).

The authors explained this result by the variation of interaction strength in the hard blocks. The accumulation of molecular force along the chain depends on the strain rate relative to the rate of disentanglement, the latter of which is strongly influenced by the interaction strength and density in the hard segment. In fact, the increase of light intensity correlated well with the increasing density of the hydrogen bonding network from PE to PA to PU. Sijbesma's study has confirmed that strong hydrogen bonding interactions can enhance the activation of mechanophores in the bulk because of the improved force transduction capability. This result agrees with the conclusions previously outlined in Section 5.3.2.1.

5.3.2.4 Mechanical Activation Enhanced by Strong Hydrogen Bonding at Chain Ends

In Weng's previous design (Section 5.3.2.1), supramolecular UPy motifs were embedded in the polymer backbone, and the supramolecular interactions between these functionalities were found to dissociate at a lower applied force than the activation of SP mechanophore. In a subsequent study,[67] Weng and coworkers investigated the interplay between supramolecular interactions from chain end UPy motifs and SP mechanophores in the chain. SP-containing prepolymers were prepared by polyurethane polymerization. The linear prepolymers were then end-capped with UPy motifs to facilitate chain association. Linear supramolecular polymers composed of alternating hard (UPy dimers) and soft (PTHF) segments were readily obtained (Figure 5.10a). To emphasize the role of supramolecular interactions, the same prepolymers were reacted with ethanol to prepare controls with much weaker inter-chain associations. The UPy-decorated prepolymers are referred to as "**U10**" and "**U20**", while the controls are denoted as "**E10**" and "**E20**". 10 and 20 are the molecular weights of the prepolymers in kDa.

Mechanical activation of SP was studied by tensile tests. **U10** and **U20** showed superb tensile properties and mechanochromism compared to **E10** and **E20**, respectively. During stretching, **U10** and **U20** exhibited a more intense

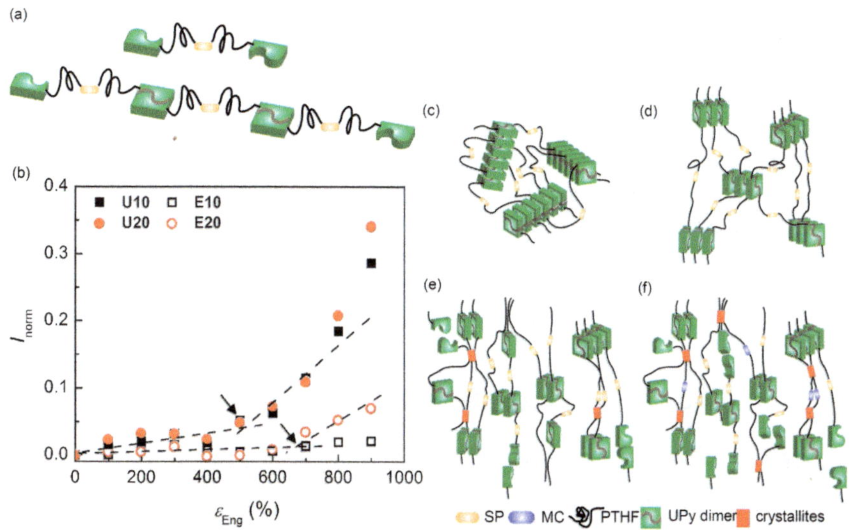

Figure 5.10 Effects of supramolecular interactions on the mechanical activation of linear supramolecular polymers. (a) Illustration of the supramolecular polymerization of SP-containing macromers. (b) The normalized fluorescent intensity as a function of strain. The black arrows indicate the onset of SP activation. (c) to (f) Proposed mechanism of mechanical activation in UPy-containing materials. Reprinted with permission from ref. 67. Copyright 2014 from American Chemical Society.

purple color in the gauge sections than the ethanol-functionalized counterparts, suggesting a higher degree of SP to MC conversion. Fluorescent spectroscopy further confirmed that observation (Figure 5.10b). MC has two characteristic emission bands at 578 and 620 nm, which were used to determine the concentration of the species in the material. From the fluorescent intensity, the ratio of *mechanically activated* SP to the total SP was calculated. This ratio was 34% for **U20** but only 7% for **E20** at a strain of 900%.

The authors interpreted the enhanced activation of SP as a result of UPy-promoted chain alignment and subsequent strain-induced crystallization (SIC). The UPy stacks aggregate and phase-separate from the PTHF matrix (Figure 5.10c), resulting in soft/hard morphology. At the beginning of deformation, the randomly orientated stacks first bear the load, but are gradually taken apart by shear stress (Figure 5.10d).[30] The soft PTHF chains then endure the load and orientate along the tensile axis. Because UPy dimers are relatively "strong", stress relaxation of the linear supramolecular chains *via* disassociation of UPy dimers is not severe. As a result, the alignment of PTHF segments at large length scale favors SIC (Figure 5.10e). The PTHF crystallites then act as additional load bearing units to accumulate sufficient molecular force to activate SP (Figure 5.10f). In contrast, the stress relaxation of the supramolecular chain in **E10** and **E20** is much faster than that in UPy materials because of the lack of strong non-covalent interactions. This in turn disfavors segment orientation as well as SIC, leading to much lower levels of activation of SP. The use of SIC to enhance the performance of stress-responsive materials is also highlighted by Sijbesma[68] and O'Bryan and McElhanon.[69]

These studies (Sections 5.3.2.1 to 5.3.2.4) for the first time demonstrate the activation of mechanophores can be enhanced by hydrogen bonding motifs incorporated in the main chain or at the chain ends if the supramolecular interactions do not restrict the mobility of the chain segments. These studies may open a new avenue to create novel stress-responsive materials whose responsiveness can be controlled by tailoring the supramolecular interactions of the system. Future work will focus on the proof-of-concept in non-crystalline polymers in both solution and bulk.

5.3.3 Effects of Van de Waals Interactions

5.3.3.1 Activation of SP in Triblock Copolymer Thermoplastic Elastomers

Although PU is facile to prepare and robust as a polymer matrix, its segmented structure and broad molecular weight distribution lead to poorly controlled hard/soft topology. On the other hand, block copolymers have found important applications as thermoplastic elastomers.[70] Their microscopic structures and consequently their mechanical properties can be rationally controlled by the monomer structure and the length of the blocks. To take advantage of these properties, Weng and coworkers introduced SP into the center of the soft block of a classical polystyrene-*b*-poly(*n*-butylacrylate)-SP-poly(*n*-butylacrylate)-*b*-polystyrene (PS-PnBA-SP-PnBA-PS) triblock copolymer (Figure 5.11a).[15] The authors fixed the length of the middle PnBA block (*ca.* 30 kDa) but varied the length of the end PS block (3, 5, 9 kDa). Samples were denoted as PS_M-$PnBA_N$-SP-$PnBA_N$-PS_M, where *N* and *M* correspond to molecular weight of PnBA and PS blocks in kDa, respectively. The dispersity of each composition of this triblock copolymer was below 1.2.

It was expected that the interaction between styrene and *n*-butylacrylate, along with the different PS block length, would lead to different microscopic structures. Both the mechanical properties and the activation of SP exhibited strong compositional dependence on the PS block. The specimen with lowest PS fraction (PS_3-$PnBA_{15}$-SP-$PnBA_{15}$-PS_3) behaved like a soft plastic with low yield strength and necking behavior during tensile drawing. Weak purple color emerged in the necking region (Figure 5.11b). Increasing the PS block to 5 kDa (PS_5-$PnBA_{15}$-SP-$PnBA_{15}$-PS_5) resulted in a characteristic elastomeric stress–strain curve and intense colorization throughout the gauge section (Figure 5.11c). The specimen with highest PS fraction (PS_9-$PnBA_{15}$-SP-$PnBA_{15}$-PS_9) showed a mechanical response of typical glassy polymers, with the highest Young's modulus but the lowest strain at break among the three. The mechanochemical reaction of SP occurred in the vicinity of the fracture point (Figure 5.11d). To exclude the effect of molecular weight, the authors also synthesized a control triblock copolymer (PS_4-$PnBA_{19}$-SP-$PnBA_{19}$-PS_4) with similar PS/PnBA ratio to that of PS_3-$PnBA_{15}$-SP-$PnBA_{15}$-PS_3 but a molecular weight in between PS_5-$PnBA_{15}$-SP-$PnBA_{15}$-PS_5 and PS_9-$PnBA_{15}$-SP-$PnBA_{15}$-PS_9. This control sample

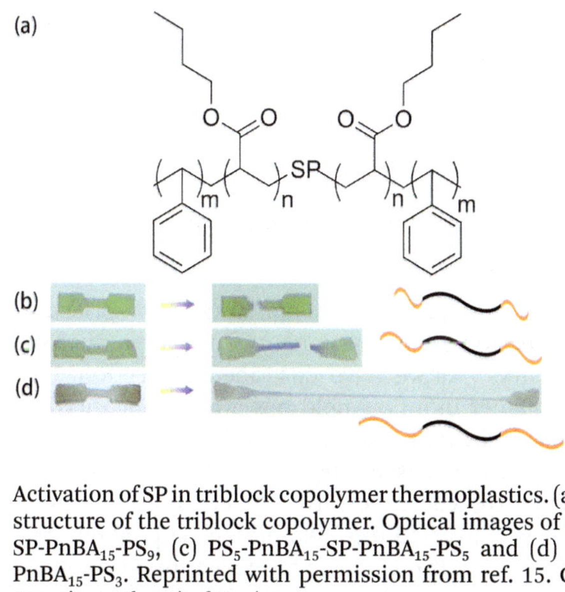

Figure 5.11 Activation of SP in triblock copolymer thermoplastics. (a) The chemical structure of the triblock copolymer. Optical images of (b) PS_9-$PnBA_{15}$-SP-$PnBA_{15}$-PS_9, (c) PS_5-$PnBA_{15}$-SP-$PnBA_{15}$-PS_5 and (d) PS_3-$PnBA_{15}$-SP-$PnBA_{15}$-PS_3. Reprinted with permission from ref. 15. Copyright 2013 American Chemical Society.

showed similar mechanical properties and activation of SP to PS_3-$PnBA_{15}$-SP-$PnBA_{15}$-PS_3. Therefore, the results were likely due to the different microscopic structures of the specimens. To obtain more evidence, transmission electron microscopy (TEM) was used to explore the microstructures. The soft/hard morphology was indeed observed in PS_5-$PnBA_{15}$-SP-$PnBA_{15}$-PS_5 and PS_9-$PnBA_{15}$-SP-$PnBA_{15}$-PS_9, but was absent in PS_3-$PnBA_{15}$-SP-$PnBA_{15}$-PS_3 and PS_4-$PnBA_{19}$-SP-$PnBA_{19}$-PS_4. Modulated differential scanning calorimetry (MDSC) and dynamic mechanical analysis (DMA) further supported this assumption.

Based on the above results, it is clear that van der Waals' interactions and the PS/PnBA ratio modulate the microphase-separation of the samples, and consequently the mechanical and mechanochromic properties. Although the chain segment in PS_3-$PnBA_{15}$-SP-$PnBA_{15}$-PS_3 has sufficient mobility (T_g of PnBA is around −40 °C), the sample lacks physical cross-links to bear the load. Viscous flow dominates the deformation of the specimen and the corresponding viscous force is not strong enough to trigger SP activation. In the case of PS_5-$PnBA_{15}$-SP-$PnBA_{15}$-PS_5, the immiscibility between the polymers leads to soft/hard morphology. The aggregated PS blocks act as effective load-bearing units and also slow down the stress relaxation of the chain. On the other hand, the PnBA blocks still have sufficient mobility to align along the tensile direction, enhancing force transfer. Hence, it showed the most pronounced activation of SP among the four samples. However, the excessive PS associations in PS_9-$PnBA_{15}$-SP-$PnBA_{15}$-PS_9 drastically reduce the mobility of PnBA segments and hence the orientation of SP. Consequently, only localized color change was seen in this material. These results with triblock copolymers with well-defined microstructures suggest that both the density of load-bearing units and the mobility of the chain segments are important to achieve comprehensive stress-responsiveness, *i.e.* the activation of SP in wide regions of the sample. Both quantities can be tuned by supramolecular interactions as well as by other environmental effects (*e.g.* temperature and the presence of solvents) as recently reported by Moore and coworkers.[71]

5.3.3.2 *Tuning Non-covalent Interactions by Polymer Mechanochemistry*

In a unique study, Craig, Mahanthappa and coworkers succeeded in altering monomer–monomer interactions and polymer microstructures using mechanochemistry.[72] As we have described in Section 5.2.1, the internal tension on a polymer in a flow field has a parabolic distribution along the chain contour with the maximum value at the middle point. For a homopolymer with multiple, non-scissile mechanophores, the segments continue to be activated from the center towards the chain ends, leading to a triblock copolymer that is converted into two diblock copolymer chains after chain scission (Figure 5.12a). If the activated mechanophores are immiscible with the non-activated mechanophores, microphase-separation may occur.

For a proof-of-concept demonstration, the authors synthesized poly(*gem*-dichlorocyclopropane) (poly(*g*DCC)). The *g*DCC mechanophore undergoes a ring–opening reaction to generate 2,3-dichloroalkene species under mechanical or thermal stimulus.[73] Sonicating a poly(*g*DCC) sample (93 kDa) under pulsed ultrasound for 30 min. led to an average of one scission per chain (*ca.* 51 kDa) and 49% conversion to 2,3-dichloroalkene. The ^{1}H NMR resonances of the vinylic and allylic hydrogens in the product showed distinct chemical shift in comparison to those in the thermolysis generated product. These deviations result from the different adjacent groups to the 2,3-dichloroalkene in "blocky" or "random" repeating segments along the chain. To further confirm the blocky nature of the activated zone, the alkene moieties were cleaved by ozonolysis to isolate non-activated *g*DCC moieties. The average molecular weight of the *g*DCC blocks (*ca.* 32 kDa) after ozonolysis coincided with the average unreacted *g*DCC content (30 kDa) in the post-sonicated product.

Because of the immiscibility between the polymer segments made up of *g*DCC and the 2,3-dichloroalkene, the diblock copolymer products showed distinct phase behavior in the bulk. DSC tests revealed two endotherms at 2 and 21 °C, corresponding to the poly(*g*DCC) and 2,3-dichloroalkene blocks, respectively. The annealed product exhibited multiple scattering peaks by SAXS, suggesting an ordered lamellar morphology with a lamellar spacing of 27 nm (Figure 5.12b).

Craig's work has demonstrated that polymer mechanochemistry is able to create unique species and alter the van der Waals interaction between monomers, and thus the condensed state of the polymer chains. Their platform offers a new way to "synthesize" block copolymers that would otherwise require harsh conditions.

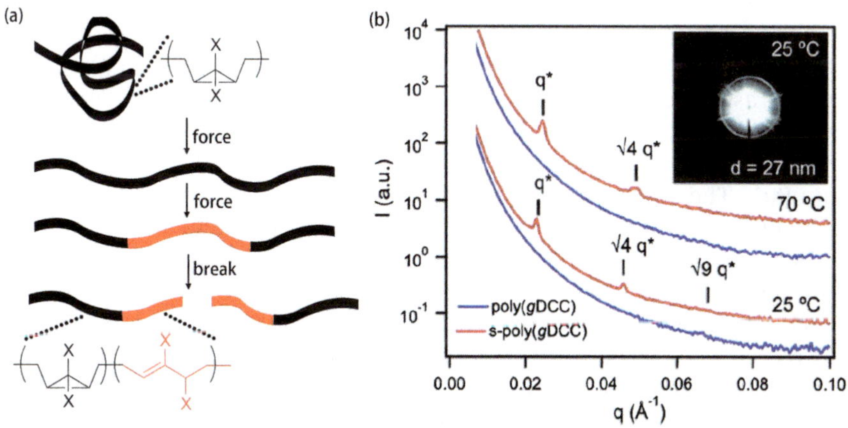

Figure 5.12 Polymer mechanochemistry changes monomer–monomer interactions and thus the microscopic structure. (a) Pulsed ultrasound of *g*DCC results in diblock copolymers. (b) SAXS of sonicated poly(*g*DCC) shows long-range order structure. Reprinted with permission from ref. 72. Copyright 2011 American Chemical Society.

5.3.4 Effect of Metallo-supramolecular Interactions

Unlike hydrogen bonding and van der Waals interactions, metallo-supramolecular interactions show several advantages: (i) the bond strengths can be fine-tuned by the choice of metal ions and ligands, ranging from weak interactions to values that are close to "classical" covalent bonds;[74] (ii) metal–ligand bonds are kinetically labile and respond to various external stimuli, accompanied by changes in optical, catalytic, electrical and chemical properties;[75–80] (iii) the metal–ligand complex is often highly charged, leading to a high potential for phase-separation in the bulk polymers. Because of these unique features, metallo-supramolecular interactions have found wide applications in self-healing materials,[81,82] biomimetic materials[83] and shape memory polymers,[84] to name a few. Herein we describe two examples of metallo-supramolecular interactions used as mechanophores or as physical cross-links to enhance the activation of covalent mechanophores.

5.3.4.1 *Using Metal–ligand Complex as Mechanophores*

Sijbesma and coworkers first highlighted the use of silver(I) or ruthenium(II) coordinated with *N*-heterocyclic carbenes (NHC) as latent catalysts. These organometallic complexes break under mechanical force, and metal–ligand complex can trigger small molecule reactions and polymerizations.[68,85]

Recently, Weder and coworkers reported mechanochemical transduction in metallo-supramolecular polymers.[86] Specifically, the authors prepared poly(ethylene-*co*-butylene) end-capped with 2,6-bis(1′-methyl-benzimidazolyl)pyridine (Mebip) ligands. Binding the Mebip ligands to the europium ions Eu^{3+} afforded a metallo-supramolecular network (Figure 5.13a). Upon ultrasound sonication of the network in dilute solution, the photoluminescence of the Eu^{3+}: Mebip complex varied from red to blue; the red color then returned within a few minutes once the sonication ceased. The phenomenon was ascribed to the disassembly and reassembly of the Eu^{3+}: Mebip complex. On the other hand, sonicating a low-molecular-weight model complex between monotopic Mebip$C_{12}H_{25}$ and Eu^{3+} resulted in no change of emission from Eu^{3+}. The authors thus concluded the ultrasound-induced dissociation of the Eu^{3+}: Mebip complex was solely driven by mechanical force.

The result inspired the authors to explore the possibility of self-healing materials that were "*mechanically*" healable. Films of this polymer were bisected, lap-jointed and sonicated (Figure 5.13b). After sonication, the two pieces were welded together, resulting in a specimen with completely restored mechanical properties.

The films not only exhibit self-healing properties, but they also participate in metal-exchange reactions regulated by mechanical activation. Iron(II) can form strong complex with Mebip and display a deep purple color. When the authors swelled the Eu^{3+}: Mebip films in $Fe(ClO_4)_2$ in CH_3CN until equilibrium (Figure 5.13c), the color of the soaked films was almost unchanged, indicating negligible metal-exchange. However, sonicating the imbibed films

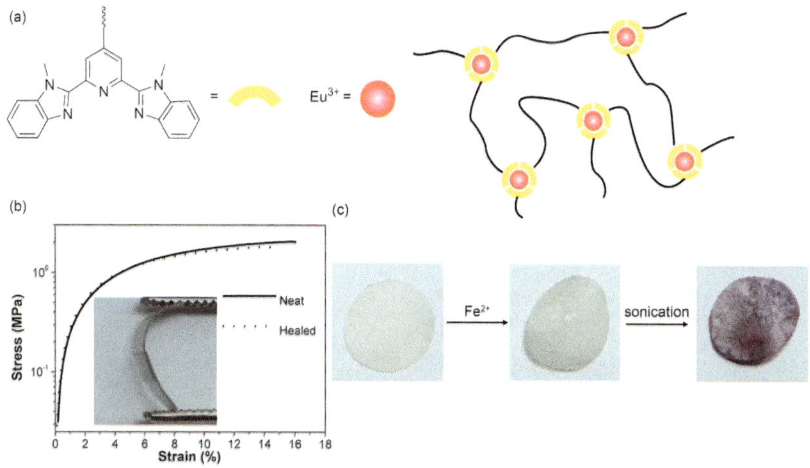

Figure 5.13 Supramolecular mechanophore based on metal–ligand interactions. (a) Chemical structure of Mebip and the formation of metallo-supramolecular polymer networks. (b) Self-healing of the metallo-supramolecular polymers triggered by mechanical force. (c) Mechanically induced metal exchange. The film was first swollen in $Fe(ClO_4)_2$ solution for 5 days and subsequently ultrasonicated in neat CH_3CN for 60 min. Reprinted with permission from ref. 86. Copyright 2014 American Chemical Society.

in neat CH_3CN for 60 min. resulted in a color change to deep purple, which was characteristic of the Fe^{2+}: Mebip complex. Weder's study underlines the dual roles of metal–ligand complexes as versatile supramolecular mechanophores and optical probes.

5.3.4.2 Effect of Metallo-supramolecular Interactions on the Activation of Mechanophores

Following their studies on the ability to enhance the activation of SP through hydrogen bonding interactions (Sections 5.3.2.1 and 5.3.2.4), Weng and coworkers next focused on utilizing metallo-supramolecular interactions. In recent years, the Weng group has harnessed BTP, a tridentate ligand that can be obtained *via* CuAAC "click chemistry" (Figure 5.14a), in making supramolecular polymers.[24,87] In a recent work, Weng and coworkers incorporated 15 BTP motifs and 0.4 SP moieties into a linear polymer backbone (32 kDa) through polyurethane polymerization.[83] Upon addition of Zn^{2+} or Eu^{3+} into the ligand macromolecules, thermally stable (up to 150 °C) hard/soft morphology was formed because of the immiscibility between polar metal: BTP complex and the apolar PTHF segment. The aggregated metal: BTP complexes served as physical cross-links. Correspondingly, the metallo-polymer films exhibited superior mechanical properties over the control film (*i.e.* no addition of metal ions).

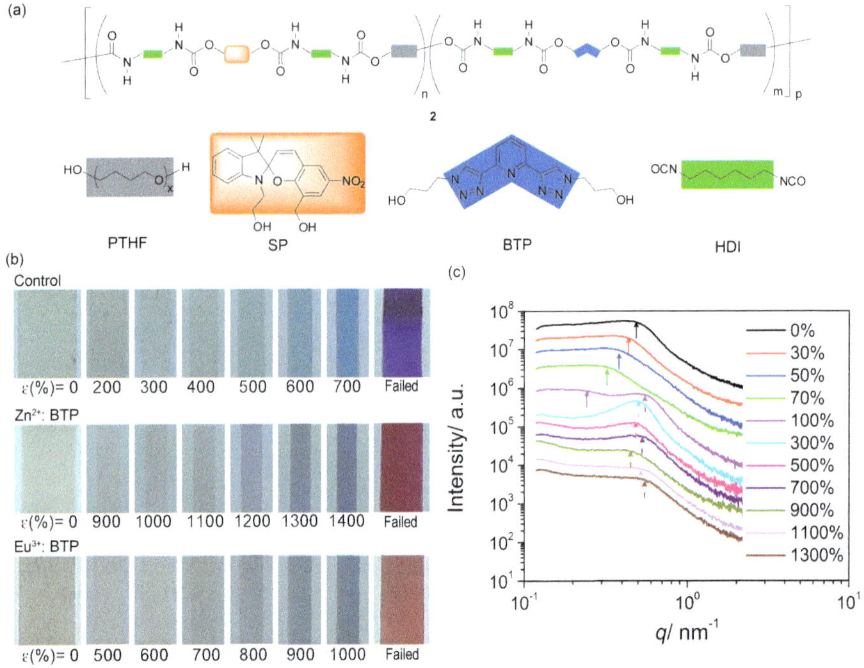

Figure 5.14 Effects of metallo-supramolecular interactions on the mechanical activation of SP. (a) Chemical structure of the metallo-polymers. (b) Force-induced color variation for the control sample and the metallo-polymer films. (c) SAXS data for a Zn^{2+}: BTP film stretched at various strains as indicated in the legend. The solid color arrows indicate the shifting of the primary peak. The dashed arrows point out the appearance of a new peak. Reprinted with permission from ref. 83. Copyright 2013 American Chemical Society.

When subject to conditions that caused fracture, the control film changed from yellow to purple, similar to the results obtained in the hydrogen bonding systems. In contrast, the metallopolymer films pre-treated by intense white light turned red after failure (Figure 5.14b). Metal ions are known to stabilize thermal isomerization of SP to MC at room temperature, and the red color is a characteristic of the metal-MC interaction.[88,89] Such results suggested a scenario of metal: BTP disassembly, mechanical activation of SP to MC and coordination between the metal ions and MC.

To gain further insights into the mechanical and mechanochromic properties of the films, the authors performed strain-dependent SAXS tests (Figure 5.14c). For a Zn^{2+}: BTP film, the primary Bragg peak (solid arrows) shifted to low q region below a strain of 100%, which reflected the increasing spacing between hard domains (Zn^{2+}: BTP aggregates). Above 100%, the Zn^{2+}: BTP aggregates were gradually broken by high stress, leading to a redistribution of the inter-domain distance, as evidenced by the appearance of another peak (dashed arrows). The secondary peak started to diminish from a strain of 300%, indicating almost all the hard domains had been fractured. On the other hand, visible color change began at a strain of 1100% for this material. Moreover, the reported force (by AFM) to break terpyridine-based metal complexes is 100 pN,[90] and is lower than that to activate SP (240–260 pN[62]). Collectively, the metal: BTP aggregate (hard domain) disassembles first, followed by the dissociation of the metal: BTP complex, and eventually the activation of SP to MC. The partially released metal ions then coordinate with MC and give rise to the red color. Again, a similar "gate" effect is expected for the metallo-supramolecular motifs as was observed in the UPy-containing system discussed in Section 5.3.2.1.[58]

5.4 Non-covalent Interactions in Stress-responsive Materials

5.4.1 Activation of Mechanophores and Irreversible Deformation of the Matrix

Although mechanophores have offered researchers great opportunities, the fabrication of stress-responsive materials with repeatable mechanochemical activation poses a great challenge. Moore and coworkers pioneered the activation of SP in the bulk with sufficient plastic deformation. They prepared glassy poly(methyl methacrylate) (PMMA) beads cross-linked by SP.[91] Compressing the beads caused strong color change. They found the yield stress coincided well with the threshold stress to activate SP. Later, more studies showed that activation events in bulk accompanied irreversible deformation or even macroscopic failure.[71,92,93] In these experiments, the shape of the activated sample was completely different from its nascent form, precluding cyclic activation of mechanophores.

The key to reversible activation is to balance the force required to trigger the mechanophores with that which causes irreversible deformation. Elastomers are known to recover their original shapes when the stress is released. Previous attempts to incorporate mechanophores into elastomers,[94,95] however, were less effective because of the inherently brittle nature of covalent cross-links. An ideal substrate should endure high stress accompanied by large deformation and recover its macroscopic shape, an area in which supramolecular interactions are powerful tools.[60,61,96–99]

5.4.2 Successive Mechanochemical Activation in Hydrogen-bonded Reinforced Elastomers

The rapid development of supramolecular chemistry has opened the door to the use of strong and tough matrices in stress-responsive materials. In that regard, supramolecular motifs can be introduced to create transient junctions. These physical junctions not only contribute to modulus and strength, but also consume strain energy and protect covalent bonds *via* dissociation, offering materials with incredible strength, toughness and resilience.

Recently, Weng's group incorporated SP into a polyurethane elastomer doubly cross-linked by both covalent and non-covalent cross-linkers.[100] The elastomer contained oligo-poly(lactic acid)-capped pentaerythritol as covalent cross-links, UPy motifs as transient junctions and SP as mechanochromic moieties. Again, the UPy dimers built long stacks, and the stacks aggregated and phase-separated with the soft PTHF segments, thus acting as additional physical cross-links.

Snapshots of the elastomer under 10 cyclic loading are shown in Figure 5.15a. In the first loading cycle, the color of the specimen changed from yellow to blue, although it soon turned to purple once the stress was released. The secondary colorization was completely reversible in the additional 9 cycles, with the blue color always changing to purple in the relaxed state. To gain deep insight into the blue-purple transition,

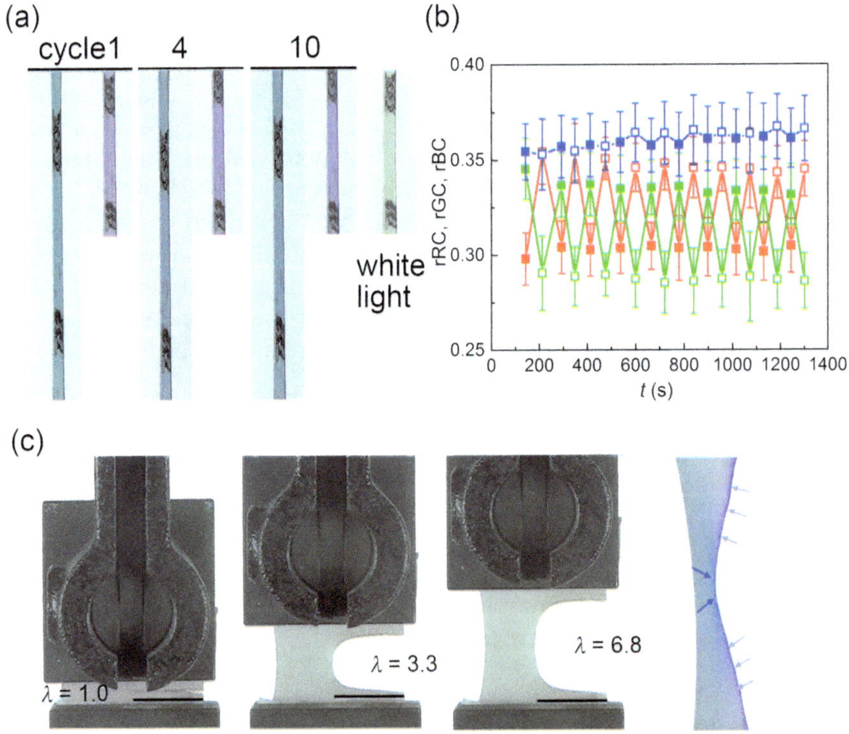

Figure 5.15 Multiple activation in a dual cross-linked elastomer. (a) Activation of SP over 10 cycles. (b) The change of the relative R, G and B signals during cyclic loading. rRC; red symbols; rGC: green symbols; rBC: blue symbols. (c) SP as a molecular force probe to map the stress distribution around a crack tip during propagation. Reprinted with permission from ref. 100. Copyright 2014 American Chemical Society.

the authors performed RGB analysis. Images were analyzed by ImageJ to extract the value of each channel (R, G and B). If SP was activated to MC, the blue part of the RGB ratio (denoted as rBC = B/(R + G + B)) was nearly constant during the 10 cycles (Figure 5.15b). However, the red (denoted as rRC = R/(R + G + B)) and green (denoted as rGC = G/(R + G + B)) parts depended very much on the stress. The maximum and minimum values of rRC were located at stress-free and strained states, respectively, whereas the rGC signal changed inversely to rRC. Accordingly, the specimen varied from blue to purple when the stress was removed. The changes of both rRC and rGC were repeatable and reversible because of the elastic nature of the substrate.

The Weng and Moore groups both noted the blue-purple change in previous studies,[58,83,91] and recognized the blue color as a transient state, *i.e.* only stable under mechanical force. Craig and coworkers first suggested that this transition is the result of an isomerization about the methane bridge of MC.[101] Weng and coworkers believe that such features give SP the capability to act as a mechanochromic probe for studying the fracture mechanism of polymers under complicated loading conditions, *e.g.* crack propagation.

For a proof-of-concept demonstration, Weng and coworkers investigated the activation of SP during crack propagation. A specimen was notched and stretched perpendicular to the direction of the crack (Figure 5.15c). As the strain approached a critical value (*ca.* 580%, $\lambda = 680\%$), a pale blue region appeared ahead of the crack tip, in which the polymer chains were highly stretched. Further stretching the specimen resulted in slow propagation of the crack tip. Two regions could be distinguished: a blue region close to and a purple region far away from the crack tip. The blue region was highly stress-concentrated and the purple region was stress-free. Therefore, the distribution of the stress can be easily visualized by the naked eye. However, it should be emphasized that the map of stress is feasible only in a tough substrate. Without the reinforcement of the UPy motifs, the activation of SP is accompanied with extremely fast crack propagation, because of the brittle nature of the covalent bonds, which complicates the detection of the cracks.

Boydston and coworkers recently reported an oxanorbornadiene mechanophore which undergoes bond-flexing to trigger retro-[4 + 2] cycloaddition and release furan.[92] The authors incorporated the mechanophore into a cross-linked poly(methyl acrylate) (PMA) network. Although they successfully demonstrated the activity of the mechanophore, the required compressive force caused macroscopic failure of the matrix. To achieve repeatable activation under multiple cyclic loading, they replaced PMA with a cross-linked polyurethane network.[93] The polyurethane samples were subjected to nine compression cycles at a compression strength of 88 MPa. The cumulative activation fraction reached a maximum of 6–7% activation. Although stress-softening was observed, the specimen could resist permanent deformation after multiple compressions.

5.5 Conclusions and Outlook

5.5.1 Conclusions

Polymer mechanochemistry is a fundamentally multiscale phenomenon. The final bond scission and the activation of mechanophores involve a series of events. The macroscopic deformation inputs strain energy to the system. To reduce the total free energy, polymer chains undergo various dynamic processes (slip, disentanglement and SIC) to relax stress. Once the strain energy is above critical value, chains are unraveled to highly extended conformations (*i.e.* coil-to-stretch or globule–stretch transitions), generating molecular strains. The molecular strains distort the molecular orbitals, redistributing the electronic density, and finally promoting/inhibiting chemical reactions. Accordingly, the effects of supramolecular interactions on polymer mechanochemistry can be understood at different length scales:

(1) Molecular length scale. Many supramolecular motifs themselves are potential mechanophores and their fragments are functional. Examples include the mechanoluminescent Eu^{3+}: Mebip complexes and the mechanocatalytic silver(I) or ruthenium(II) NHC complexes.[85,86]
(2) Segment length scale. Intra-chain supramolecular interactions promote the folding of the chains. Two effects can be noted: (i) the folded domains may preserve weak covalent bonds or mechanophores from solvent or attacking agents, improving their chemical stability (*e.g.* the disulfide bonds in VCAM-1 [53]); (ii) the folded domains also provide additional hydrodynamic shielding for the inner segments. The whole chain therefore experiences much less hydrodynamic drag force than a random coil.
(3) Chain length scale. Mechanochemical reactions entail chain extension to generate sufficient molecular strains. We have noticed: (i) the dissociation of the supramolecular moieties consumes a great amount of strain energy hence delays the chain extension process; (ii) supramolecular interactions may alter the chain topology under extended state (loop or ladder-like polymers), reducing the maximum contour length exposed to the flow field and hence reducing the internal tension (eqn (5.1)).
(4) Condensed state. (i) The strong association of the supramolecular moieties slows down the stress relaxation of the polymer chains. (ii) The supramolecular moieties may lead to soft/hard morphology (additional physical cross-links). The above two effects facilitate force transduction to the mechanophores therefore enhance their activation. (iii) Excessive supramolecular interactions drastically slow down the dynamics of the chains. In this case, the constrained segments hinder force transmission to nearby regions. As a result, only localized activation of mechanophore can be observed.

5.5.2 Outlook

Studies by Weng's group and several other research groups have demonstrated that supramolecular interactions are powerful tools to tailor the reactivity of covalent bonds and the response of stress-responsive materials. However, scientists are still at the early stage in this area and the following prospects may deserve more analysis.

(1) Decoupling the effects of supramolecular interactions. In dilute solution, supramolecular interactions mainly play three roles in polymer mechanochemistry. Can we estimate the relative importance of each mechanism *via* well-defined polymer architecture? For example, if a ring chain formed by supramolecular binding (Figure 5.2d) shows lower mechanochemical reactivity than a linear counterpart having the same molecular weight, the change of chain topology should play a dominant role compared to the other mechanisms.
(2) Using supramolecular interactions for advanced design of mechanophores. Inspired by the thiol/disulfide exchange in stretched proteins, we expect more supramolecular "gate" designs to tune the mechanochemical reactivity of covalent mechanophores. Synthetic approaches to these structures and their use in quantitative characterizations represent great challenges in the field.
(3) Effect of hierarchical structure on the polymer mechanochemistry. In this chapter, most of the stress-responsive materials have simple hard/soft morphology at microscopic length scale. Conversely, Nature utilizes supramolecular interactions to construct hierarchical structures for functional biomaterials.[102] Do other microscopic structures (*e.g.* lamellar) enhance the activation of mechanophores more effectively than the hard/soft morphology? What is the best place to incorporate mechanophores in these structures? Answers to these questions may lead to novel stress-responsive materials with comprehensive performances.
(4) Remote control of supramolecular interactions and microscopic structures by polymer mechanochemistry. Craig's remarkable example[101] (Section 5.3.2.2) provides a potential method to alter monomer–monomer interactions, and hence the microscopic structure, by mechanical force. However, more examples are required to demonstrate its versatility.

Acknowledgements

This work was supported by the National Natural Science Foundation of China (No. 21304076 and No. 21574108) and China Postdoctoral Science Foundation (No. 2013M541857 and No. 2014T70612).

References

1. C. L. Brown and S. L. Craig, Molecular engineering of mechanophore activity for stress-responsive polymeric materials, *Chem. Sci.*, 2015, **6**(4), 2158–2165.
2. J. N. Brantley, K. M. Wiggins and C. W. Bielawski, Polymer mechanochemistry: the design and study of mechanophores, *Polym. Int.*, 2013, **62**(1), 2–12.
3. L. Takacs, The historical development of mechanochemistry, *Chem. Soc. Rev.*, 2013, **42**(18), 7649–7659.
4. M. K. Beyer and H. Clausen-Schaumann, Mechanochemistry: The Mechanical Activation of Covalent Bonds, *Chem. Rev.*, 2005, **105**(8), 2921–2948.
5. K. M. Wiggins, J. N. Brantley and C. W. Bielawski, Polymer Mechanochemistry: Force Enabled Transformations, *ACS Macro Lett.*, 2012, **1**(5), 623–626.
6. P. A. May and J. S. Moore, Polymer mechanochemistry: techniques to generate molecular force *via* elongational flows, *Chem. Soc. Rev.*, 2013, **42**(18), 7497–7506.
7. M. M. Caruso, D. A. Davis, Q. Shen, S. A. Odom, N. R. Sottos and S. R. White, *et al.*, Mechanically-Induced Chemical Changes in Polymeric Materials, *Chem. Rev.*, 2009, **109**, 5755–5798.
8. H. Zhang, Y. Lin, Y. Xu and W. Weng, Mechanochemistry of Topological Complex Polymer Systems, in *Polymer Mechanochemistry*, ed. Boulatov R., Topics in Current Chemistry, Springer International Publishing, Berlin, 2015, vol. 369, pp. 135–207.
9. P. G. De Gennes, Coil-stretch transition of dilute flexible polymers under ultrahigh velocity gradients, *J. Chem. Phys.*, 1974, **60**(12), 5030–5042.
10. A. Keller and J. A. Odell, The extensibility of macromolecules in solution-a new focus for macromolecular science, *Colloid Polym. Sci.*, 1985, **263**(3), 181–201.
11. J. A. Odell and A. Keller, Flow-induced chain fracture of isolated linear macromolecules in solution, *J. Polym. Sci., Part B: Polym. Phys.*, 1986, **24**(9), 1889–1916.
12. R. Groote, B. M. Szyja, E. A. Pidko, E. J. M. Hensen and R. P. Sijbesma, Unfolding and Mechanochemical Scission of Supramolecular Polymers Containing a Metal–Ligand Coordination Bond, *Macromolecules*, 2011, **44**(23), 9187–9195.
13. J. A. Odell, A. J. Muller, K. A. Narh and A. Keller, Degradation of polymer-solutions in extensional flows, *Macromolecules*, 1990, **23**(12), 3092–3103.
14. C. R. Hickenboth, J. S. Moore, S. R. White, N. R. Sottos, J. Baudry and S. R. Wilson, Biasing reaction pathways with mechanical force, *Nature*, 2007, **446**(22), 423–427.
15. S. Jiang, L. Zhang, T. Xie, Y. Lin, H. Zhang and Y. Xu, *et al.*, Mechanoresponsive PS-PnBA-PS Triblock Copolymers *via* Covalently Embedding Mechanophore, *ACS Macro Lett.*, 2013, **2**, 705–709.

16. A. Alexander-Katz, M. F. Schneider, S. W. Schneider, A. Wixforth and R. R. Netz, Shear-Flow-Induced Unfolding of Polymeric Globules, *Phys. Rev. Lett.*, 2006, **97**(13), 138101.
17. C. E. Sing and A. Alexander-Katz, Giant Nonmonotonic Stretching Response of a Self-Associating Polymer in Shear Flow, *Phys. Rev. Lett.*, 2011, **107**(19), 198302.
18. C. E. Sing and A. Alexander-Katz, Globule–Stretch Transitions of Collapsed Polymers in Elongational Flow Fields, *Macromolecules*, 2010, **43**(7), 3532–3541.
19. A. Alexander-Katz and R. R. Netz, Dynamics and Instabilities of Collapsed Polymers in Shear Flow, *Macromolecules*, 2008, **41**(9), 3363–3374.
20. T. Xiao, X. Feng, S. Ye, Y. Guan, S.-L. Li and Q. Wang, *et al.*, Highly Controllable Ring-Chain Equilibrium in Quadruply Hydrogen Bonded Supramolecular Polymers, *Macromolecules*, 2012, **45**(24), 9585–9594.
21. J. G. H. Cifre, R. Pamies, M. C. L. Martinez and J. G. de la Torre, Steady-state behavior of ring polymers in dilute flowing solutions *via* Brownian dynamics, *Polymer*, 2005, **46**(1), 267–274.
22. R. P. Sijbesma, F. H. Beijer, L. Brunsveld, B. J. B. Folmer, J. H. K. K. Hirschberg and R. F. M. Lange, *et al.*, Reversible Polymers Formed from Self-Complementary Monomers Using Quadruple Hydrogen Bonding, *Science*, 1997, **278**, 1601–1604.
23. H. Kautz, D. J. M. Van Beek, R. P. Sijbesma and E. W. Meijer, Cooperative End-to-End and Lateral Hydrogen-Bonding Motifs in Supramolecular Thermoplastic Elastomers, *Macromolecules*, 2006, **39**(13), 4265–4267.
24. W. Weng, X. Fang, H. Zhang, H. Peng, Y. Lin and Y. Chen, Multiresponsive supramolecular gels constructed by orthogonal metal–ligand coordination and hydrogen bonding, *Eur. Polym. J.*, 2013, **49**(12), 4062–4071.
25. X. Yan, F. Wang, B. Zheng and F. Huang, Stimuli-responsive supramolecular polymeric materials, *Chem. Soc. Rev.*, 2012, **41**, 6042–6065.
26. E. A. Fogleman, W. C. Yount, J. Xu and S. L. Craig, Modular, Well-Behaved Reversible Polymers from DNA-Based Monomers, *Angew. Chem., Int. Ed.*, 2002, **41**(21), 4026–4028.
27. U. S. Agarwal and R. A. Mashelkar, Hydrodynamic shielding induced stability of zipping macromolecules in enlongational flows, *J. Chem. Phys.*, 1994, **100**(8), 6055–6061.
28. W. P. J. Appel, G. Portale, E. Wisse, P. Y. W. Dankers and E. W. Meijer, Aggregation of Ureido-Pyrimidinone Supramolecular Thermoplastic Elastomers into Nanofibers: A Kinetic Analysis, *Macromolecules*, 2011, **44**, 6776–6784.
29. R. M. Versteegen, R. Kleppinger, R. P. Sijbesma and E. W. Meijer, Properties and Morphology of Segmented Copoly(ether urea)s with Uniform Hard Segments, *Macromolecules*, 2005, **39**(2), 772–783.
30. E. Wisse, A. J. H. Spiering, F. Pfeifer, G. Portale, H. W. Siesler and E. W. Meijer, Segmental Orientation in Well-Defined Thermoplastic Elastomers Containing Supramolecular Fillers, *Macromolecules*, 2009, **42**(2), 524–530.

31. R. H. J. Sellin, J. W. Hoyt, J. Pollert and O. Scrivener, The effect of drag reducing additives on fluid-flows and their industrial applications. 2. present applications and future proposals, *J. Hydraul. Res.*, 1982, **20**(3), 235–292.
32. P. R. Kenis, Turbulent flow friction reduction effectiveness and hydrodynamic degradation of polysaccharides and synthetic polymers, *J. Appl. Polym. Sci.*, 1971, **15**(3), 607–618.
33. A. G. Fabula, J. W. Hoyt and H. R. Crawford, On the relevance of intrinsic viscosity to the concentration dependence of the toms effect, *Bull. Am. Phys. Soc.*, 1963, **8**, 3662–3663.
34. W. Brostow, Drag reduction in flow: Review of applications, mechanism and prediction, *J. Ind. Eng. Chem.*, 2008, **14**(4), 409–416.
35. O. Kim, R. Little, R. Patterson and R. Ting, Polymer structures and turbulent shear stability of drag reducing solutions, *Nature*, 1974, **250**, 408–410.
36. C. C. Gryte, C. Koroneos, A. Agarwal and A. Hochberg, Drag reduction characteristics of graft co-polymers prepared by the gamma-irradiation of poly(oxyethylene) in the presence of acrylic acid, *Polym. Eng. Sci.*, 1980, **20**(7), 478–484.
37. S. R. Deshmukh and R. P. Singh, Drag reduction characteristics of graft-copolymers of xanthangum and polycaylamide, *J. Appl. Polym. Sci.*, 1986, **32**(8), 6163–6176.
38. E. Dschagarowa and T. Bochossian, Drag reduction in polymer mixtures, *Rheol. Acta*, 1978, **17**(4), 426–432.
39. G. V. Reddy and R. P. Singh, Drag reduction effectiveness and shear stability of polymer-polymer and polymer-fiber mixtures in recirculatory turbulent-flow of water, *Rheol. Acta*, 1985, **24**(3), 296–311.
40. J. P. Malhotra, P. N. Chaturvedi and R. P. Singh, Drag reduction by polymer polymer mixtures, *J. Appl. Polym. Sci.*, 1988, **36**(4), 837–858.
41. J. P. Malhotra, S. R. Deshmukh and R. P. Singh, Turbulent drag reduction bu polymer-fiber mixtures, *J. Appl. Polym. Sci.*, 1987, **33**(7), 2467–2478.
42. R. Y. Ting and R. C. Little, Shear Stability of drag-reducing Polyacrylic Acid Solutions, *Nat. Phys. Sci.*, 1973, **241**, 42–44.
43. R. M. Kowalik, I. Duvdevani, D. G. Peiffer, R. D. Lundberg, K. Kitano and D. N. Schulz, Enhanced drag reduction *via* interpolymer associations, *J. Non-Newtonian Fluid Mech.*, 1987, **24**(1), 1–10.
44. S. Malik, S. N. Shintre and R. A. Mashelkar, Enhancing the shear stability in drag-reducing polymers through molecular associations, *Macromolecules*, 1993, **26**(1), 55–59.
45. S. Malik and R. A. Mashelkar, Hydrogen-bonding mediated shear stable clusters as drag reducers, *Chem. Eng. Sci.*, 1995, **50**(1), 105–116.
46. V. Vogel, Mechanotransduction involving multimodular proteins: Converting force into biochemical signals, *Annu. Rev. Biophys. Biomol. Struct.*, 2006, **35**, 459–488.
47. E. Evans, Probing the Relation Between Force—Lifetime—and Chemistry in Single Molecular Bonds, *Annu. Rev. Biophys. Biomol. Struct.*, 2001, **30**(1), 105–128.

48. I. B. Bekard, K. J. Barnham, L. R. White and D. E. Dunstan, alpha-Helix unfolding in simple shear flow, *Soft Matter*, 2011, **7**(1), 203–210.
49. E. Paci and M. Karplus, Forced unfolding of fibronectin type 3 modules: an analysis by biased molecular dynamics simulations1, *J. Mol. Biol.*, 1999, **288**(3), 441–459.
50. P. M. Williams, S. B. Fowler, R. B. Best, J. Luis Toca-Herrera, K. A. Scott and A. Steward, *et al.*, Hidden complexity in the mechanical properties of titin, *Nature*, 2003, **422**(6930), 446–449.
51. F. Gräter, J. Shen, H. Jiang, M. Gautel and H. Grubmüller, Mechanically Induced Titin Kinase Activation Studied by Force-Probe Molecular Dynamics Simulations, *Biophys. J.*, 2005, **88**(2), 790–804.
52. N. Bruns, K. Pustelny, L. M. Bergeron, T. A. Whitehead and D. S. Clark, Mechanical Nanosensor Based on FRET within a Thermosome: Damage-Reporting Polymeric Materials, *Angew. Chem., Int. Ed.*, 2009, **48**(31), 5666–5669.
53. N. Bhasin, P. Carl, S. Harper, G. Feng, H. Lu and D. W. Speicher, *et al.*, Chemistry on a Single Protein, Vascular Cell Adhesion Molecule-1, during Forced Unfolding, *J. Biol. Chem.*, 2004, **279**(44), 45865–45874.
54. S. R. Koti Ainavarapu, A. P. Wiita, L. Dougan, E. Uggerud and J. M. Fernandez, Single-Molecule Force Spectroscopy Measurements of Bond Elongation during a Bimolecular Reaction, *J. Am. Chem. Soc.*, 2008, **130**(20), 6479–6487.
55. S. Garcia-Manyes, J. Liang, R. Szoszkiewicz, T.-L. Kuo and J. M. Fernández, Force-activated reactivity switch in a bimolecular chemical reaction, *Nat. Chem.*, 2009, **1**(3), 236–242.
56. T. J. Kucharski, Z. Huang, Q.-Z. Yang, Y. Tian, N. C. Rubin and C. D. Concepcion, *et al.*, Kinetics of Thiol/Disulfide Exchange Correlate Weakly with the Restoring Force in the Disulfide Moiety, *Angew. Chem., Int. Ed.*, 2009, **121**(38), 7174–7177.
57. Q. Z. Yang, Z. Huang, T. J. Kucharski, D. Khvostichenko, J. Chen and R. Boulatov, A molecular force probe, *Nat. Nanotechnol.*, 2009, **4**, 302–306.
58. X. Fang, H. Zhang, Y. Chen, Y. Lin, Y. Xu and W. Weng, Biomimetic Modular Polymer with Tough and Stress Sensing Properties, *Macromolecules*, 2013, **46**(16), 6566–6574.
59. S. H. M. Söntjens, R. P. Sijbesma, M. H. P. van Genderen and E. W. Meijer, Stability and Lifetime of Quadruply Hydrogen Bonded 2-Ureido-4[1*H*]-pyrimidinone Dimers, *J. Am. Chem. Soc.*, 2000, **122**, 7487–7493.
60. A. M. Kushner, J. D. Vossler, G. A. Williams and Z. Guan, A Biomimetic Modular Polymer with Tough and Adaptive Properties, *J. Am. Chem. Soc.*, 2009, **131**(25), 8766–8768.
61. Z. B. Guan, J. T. Roland, J. Z. Bai, S. X. Ma, T. M. McIntire and M. Nguyen, Modular domain structure: A biomimetic strategy for advanced polymeric materials, *J. Am. Chem. Soc.*, 2004, **126**(7), 2058–2065.
62. G. R. Gossweiler, T. B. Kouznetsova and S. L. Craig, Force-Rate Characterization of Two Spiropyran-Based Molecular Force Probes, *J. Am. Chem. Soc.*, 2015, **137**(19), 6148–6151.

63. C. K. Lee, B. A. Beiermann, M. N. Silberstein, J. Wang, J. S. Moore and N. R. Sottos, *et al.*, Exploiting Force Sensitive Spiropyrans as Molecular Level Probes, *Macromolecules*, 2013, **46**(10), 3746–3752.
64. J. Wang, T. B. Kouznetsova, Z. Niu, M. T. Ong, H. M. Klukovich and A. L. Rheingold, *et al.*, Inducing and quantifying forbidden reactivity with single-molecule polymer mechanochemistry, *Nat. Chem.*, 2015, **7**, 323–327.
65. Y. Chen, A. J. H. Spiering, S. Karthikeyan, G. W. M. Peters, E. W. Meijer and R. P. Sijbesma, Mechanically induced chemiluminescence from polymers incorporating a 1,2-dioxetane unit in the main chain, *Nat. Chem.*, 2012, **4**, 559–562.
66. Y. Chen and R. P. Sijbesma, Dioxetanes as Mechanoluminescent Probes in Thermoplastic Elastomers, *Macromolecules*, 2014, **47**(12), 3797–3805.
67. Y. Chen, H. Zhang, X. Fang, Y. Lin, Y. Xu and W. Weng, Mechanical Activation of Mechanophore Enhanced by Strong Hydrogen Bonding Interactions, *ACS Macro Lett.*, 2014, **3**(2), 141–145.
68. R. T. M. Jakobs, S. Ma and R. P. Sijbesma, Mechanocatalytic Polymerization and Cross-Linking in a Polymeric Matrix, *ACS Macro Lett.*, 2013, **2**(7), 613–616.
69. G. O'Bryan, B. M. Wong and J. R. McElhanon, Stress Sensing in Polycaprolactone Films *via* an Embedded Photochromic Compound, *ACS Appl. Mater. Interfaces*, 2010, **2**(6), 1594–1600.
70. A.-V. Ruzette and L. Leibler, Block copolymers in tomorrow's plastics, *Nat. Mater.*, 2005, **4**(1), 19–31.
71. B. A. Beiermann, D. A. Davis, S. L. B. Kramer, J. S. Moore, N. R. Sottos and S. R. White, Environmental effects on mechanochemical activation of spiropyran in linear PMMA, *J. Mater. Chem.*, 2011, **21**(23), 8443–8447.
72. A. L. Black Ramirez, J. W. Ogle, A. L. Schmitt, J. M. Lenhardt, M. P. Cashion and M. K. Mahanthappa, *et al.*, Microstructure of Copolymers Formed by the Reagentless, Mechanochemical Remodeling of Homopolymers *via* Pulsed Ultrasound, *ACS Macro Lett.*, 2011, **1**(1), 23–27.
73. J. M. Lenhardt, A. L. Black and S. L. Craig, *gem*-Dichlorocyclopropanes as Abundant and Efficient Mechanophores in Polybutadiene Copolymers under Mechanical Stress, *J. Am. Chem. Soc.*, 2009, **131**, 10818–10819.
74. E. Degtyar, M. J. Harrington, Y. Politi and P. Fratzl, The Mechanical Role of Metal Ions in Biogenic Protein-Based Materials, *Angew. Chem., Int. Ed.*, 2014, **53**(45), 12026–12044.
75. Y. Zhao, J. B. Beck, S. J. Rowan and A. M. Jamieson, Rheological Behavior of Shear-Responsive Metallo-Supramolecular Gels, *Macromolecules*, 2004, **37**(10), 3529–3531.
76. W. Weng, J. B. Beck, A. M. Jamieson and S. J. Rowan, Understanding the mechanism of gelation and stimuli-responsive nature of a class of metallo-supramolecular gels, *J. Am. Chem. Soc.*, 2006, **128**(35), 11663–11672.
77. M. Burnworth, J. D. Mendez, M. Schroetert, S. J. Rowan and C. Weder, Decoupling optical properties in metallo-supramolecular poly(p-phenylene ethynylene)s, *Macromolecules*, 2008, **41**(6), 2157–2163.

78. J. R. Capadona, K. Shanmuganathan, D. J. Tyler, S. J. Rowan and C. Weder, Stimuli-responsive polymer nanocomposites inspired by the sea cucumber dermis, *Science*, 2008, **319**(5868), 1370–1374.
79. W. Weng, Z. Li, A. M. Jamieson and S. J. Rowan, Control of Gel Morphology and Properties of a Class of Metallo-Supramolecular Polymers by Good/Poor Solvent Environments, *Macromolecules*, 2009, **42**(1), 236–246.
80. W. Weng, Z. Li, A. M. Jamieson and S. J. Rowan, Effect of monomer structure on the gelation of a class of metallo-supramolecular polymers, *Soft Matter*, 2009, **5**(23), 4647–4657.
81. M. Burnworth, L. Tang, J. R. Kumpfer, A. J. Duncan, F. L. Beyer and G. L. Fiore, *et al.*, Optically healable supramolecular polymers, *Nature*, 2011, **472**(7343), 334–338.
82. J. Fox, J. J. Wie, B. W. Greenland, S. Burattini, W. Hayes and H. M. Colquhoun, *et al.* High-Strength, Healable, Supramolecular Polymer Nanocomposites, *J. Am. Chem. Soc.*, 2012, **134**(11), 5362–5368.
83. G. Hong, H. Zhang, Y. Lin, Y. Chen, Y. Xu and W. Weng, *et al.*, Mechanoresponsive Healable Metallosupramolecular Polymers, *Macromolecules*, 2013, **46**(21), 8649–8656.
84. J. R. Kumpfer and S. J. Rowan, Thermo-, Photo-, and Chemo-Responsive Shape-Memory Properties from Photo-Cross-Linked Metallo-Supramolecular Polymers, *J. Am. Chem. Soc.*, 2011, **133**(32), 12866–12874.
85. R. Groote, R. T. M. Jakobs and R. P. Sijbesma, Mechanocatalysis: forcing latent catalysts into action, *Polym. Chem.*, 2013, **4**(18), 4846–4859.
86. D. W. R. Balkenende, S. Coulibaly, S. Balog, Y. C. Simon, G. L. Fiore and C. Weder, Mechanochemistry with Metallosupramolecular Polymers, *J. Am. Chem. Soc.*, 2014, **136**(29), 10493–10498.
87. J. Yuan, X. Fang, L. Zhang, G. Hong, Y. Lin and Q. Zheng, *et al.*, Multiresponsive self-healing metallo-supramolecular gels based on "click" ligand, *J. Mater. Chem.*, 2012, **22**(23), 11515–11522.
88. J. P. Phillips, A. Mueller and F. Przystal, Photochromic Chelating Agents, *J. Am. Chem. Soc.*, 1965, **87**(17), 4020.
89. J. D. Winkler, C. M. Bowen and V. Michelet, Photodynamic Fluorescent Metal Ion Sensors with Parts per Billion Sensitivity, *J. Am. Chem. Soc.*, 1998, **120**(13), 3237–3242.
90. M. Kudera, C. Eschbaumer, H. E. Gaub and U. S. Schubert, Analysis of Metallo-Supramolecular Systems Using Single-Molecule Force Spectroscopy, *Adv. Funct. Mater.*, 2003, **13**(8), 615–620.
91. D. A. Davis, A. Hamilton, J. Yang, L. D. Cremar, D. V. Gough and S. L. Potisek, *et al.*, Force-induced activation of covalent bonds in mechanoresponsive polymeric materials, *Nature*, 2009, **459**(7243), 68–72.
92. M. B. Larsen and A. J. Boydston, “Flex-Activated” Mechanophores: Using Polymer Mechanochemistry To Direct Bond Bending Activation, *J. Am. Chem. Soc.*, 2013, **135**(22), 8189–8192.
93. M. B. Larsen and A. J. Boydston, Successive Mechanochemical Activation and Small Molecule Release in an Elastomeric Material, *J. Am. Chem. Soc.*, 2014, **136**(4), 1276–1279.

94. C. M. Kingsbury, P. A. May, D. A. Davis, S. R. White, J. S. Moore and N. R. Sottos, Shear activation of mechanophore-crosslinked polymers, *J. Mater. Chem.*, 2011, **21**(23), 8381–8388.
95. C. M. Degen, P. A. May, J. S. Moore, S. R. White and N. R. Sottos, Time-Dependent Mechanochemical Response of SP-Cross-Linked PMMA, *Macromolecules*, 2013, **46**(22), 8917–8921.
96. A. M. Kushner and Z. Guan, Modular Design in Natural and Biomimetic Soft Materials, *Angew. Chem., Int. Ed.*, 2011, **50**(39), 9026–9057.
97. Y. Chen and Z. Guan, Bioinspired Modular Synthesis of Elastin-Mimic Polymers To Probe the Mechanism of Elastin Elasticity, *J. Am. Chem. Soc.*, 2010, **132**(13), 4577–4579.
98. A. M. Kushner, V. Gabuchian, E. G. Johnson and Z. Guan, Biomimetic design of reversibly unfolding cross-linker to enhance mechanical properties of 3D network polymers, *J. Am. Chem. Soc.*, 2007, **129**(46), 14110–14111.
99. Z. Guan, Supramolecular design in biopolymers and biomimetic polymers for properties, *Polym. Int.*, 2007, **56**(4), 467–473.
100. H. Zhang, Y. Chen, Y. Lin, X. Fang, Y. Xu and Y. Ruan, *et al.*, Spiropyran as a Mechanochromic Probe in Dual Cross-Linked Elastomers, *Macromolecules*, 2014, **47**(19), 6783–6790.
101. G. R. Gossweiler, G. B. Hewage, G. Soriano, Q. Wang, G. W. Welshofer and X. Zhao, *et al.*, Mechanochemical Activation of Covalent Bonds in Polymers with Full and Repeatable Macroscopic Shape Recovery, *ACS Macro Lett.*, 2014, **3**, 216–219.
102. M. A. Meyers, J. McKittrick and P.-Y. Chen, Structural Biological Materials: Critical Mechanics-Materials Connections, *Science*, 2013, **339**(6121), 773–779.

CHAPTER 6

Mechanochemistry of Polymer Brushes

NARIYE CAVUSOGLU ATAMAN[a], JAN GENZER[b] AND HARM-ANTON KLOK[a]*

[a]Institut des Matériaux et Institut des Sciences et Ingénierie Chimiques, Laboratoire des Polymères, École Polytechnique Fédérale de Lausanne (EPFL), Bâtiment MXD, Station 12, CH-1015 Lausanne, Switzerland; [b]Department of Chemical & Biomolecular Engineering, North Carolina State University, Raleigh, North Carolina 27695, United States
*E-mail: harm-anton.klok@epfl.ch

6.1 Introduction

Surface Initiated Polymerization (SIP) was reported for the first time in 1973 by Hamann *et al.*[1] Amorphous silica surfaces that featured different types of diazo initiators were used for radical polymerization of styrene, acrylamide, acrylonitrile, acrylic acid, methyl methacrylate and vinyl pyridine. In bulk polymerization, reactions with up to 80 wt% covalently bound polymer material were obtained. In 1986, Otsu *et al.* reported for the first time the use of a living radical polymerization technique to grow polymers from a solid substrate using polystyrene-gel (PSG) beads modified with *N*,*N*-diethyldithiocarbamate as a photoiniferter. The authors prepared various diblock and triblock copolymer brushes.[2] Since then, surface-initiated controlled radical polymerization (SI-CRP) techniques have been routinely employed for the synthesis of grafted polymer coatings, also referred to as polymer brushes,

Polymer Chemistry Series No. 26
Mechanochemistry in Materials
Edited by Yoan C. Simon and Stephen L. Craig

Published by the Royal Society of Chemistry, www.rsc.org

which have attracted interest for applications in biomedicine, nanotechnology, bioelectronics, optics and many other fields.[3–6]

Polymer brushes represent ultrathin coatings composed of polymer chains that are anchored with one chain end to an interface. Strictly speaking, the term polymer brush refers to surface-tethered polymer films in which the distance between two neighboring chains is smaller than the radius of gyration of the corresponding free polymer chains in solution. As a consequence, neighboring chains sterically repel each other and extend along the direction perpendicular to the grafting surface thus adopting the brush conformation.[7–12]

Polymer brushes are prepared primarily *via* two methods, which are referred to as the "grafting-onto" and "grafting-from" strategies (Figure 6.1). Although the "grafting-onto" approach is a simple method it suffers from limited grafting density and film thickness of the brushes that can be obtained. The "grafting-from" method involves a simple synthetic protocol, which facilitates the formation of polymer brush films with high grafting densities and film thicknesses. Densely grafted polymer brushes have been shown to possess unique properties including lubrication[13,14] and low friction[15] as well as the ability to efficiently prevent non-specific protein adsorption and cell adhesion.[16–19]

Due to the covalent link between the substrate and the surface-tethered polymer chains, polymer brush films prepared *via* SIP are generally considered stable as compared to brushes assembled using non-covalent interactions.[20] However, over the past years, a number of reports have been published in which the degrafting/detachment of densely grafted, covalently surface-anchored hydrophilic polymer chains exposed to aqueous media is described. While initially not completely understood, the collective results of these studies seem to point to a mechanochemical activation of surface-tethered polymer chains, due to chain stretching of densely grafted polymer chains. The aim of this chapter is to summarize the results of these literature reports and present a phenomenological description of the mechanochemistry of surface grafted polymer brushes.

6.2 Degrafting of Polymer Brushes

Table 6.1 provides an overview of the different polymer brushes that have been reported to undergo degrafting from solid substrates. Degrafting has been observed for a broad variety of chemically different brushes and from a

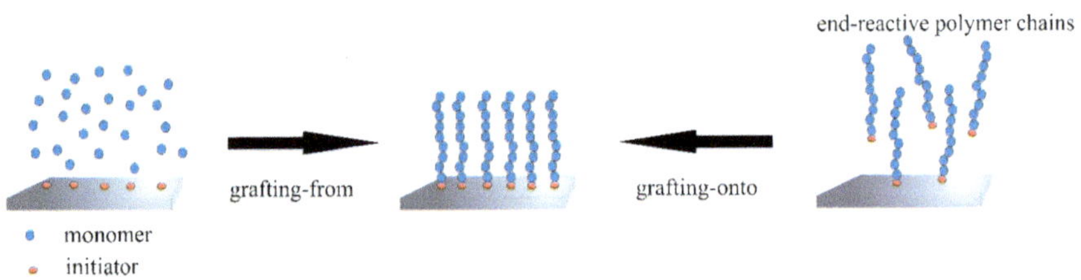

Figure 6.1 Cartoon illustrating the "grafting-from" and "grafting-onto" strategies for the preparation of polymer brushes.

Table 6.1 Overview of polymer brushes that have been reported to undergo degrafting.

Entry	Polymer brush		Substrate	Polymerization method	Incubation conditions	Remarks
1 [21]	Hyper-branched polyglycidol (HPG)	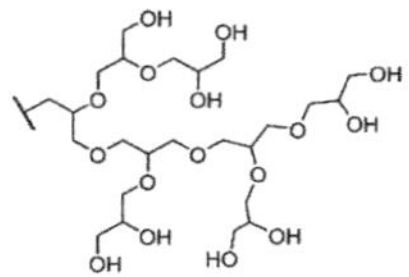	Au	AROP	Reaction conditions 40 °C	Hyperbranched polyglycidol (HPG) was grafted from Au surfaces by AROP and analyzed by AFM. After 7.5 min. Reaction time, Au surfaces showed an increase in the HPG domain size and coverage of the Au surfaces. After 20 min. Reaction time, analysis showed bare Au substrates indicating *in-situ* degrafting of the surface-grafted polymers.
2 [22]	$P(PEGMA)_{10}$	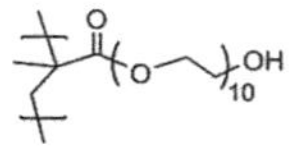	Glass, SiO_x	SI-ATRP	Cell culture medium 37 °C	$P(PEGMA)_{10}$ brushes were synthesized *via* SI-ATRP from trimethoxysilane ATRP initiator modified SiO_x surfaces. Incubation of the brushes in cell culture medium resulted in wrinkling and delamination of polymer layers. Decreasing the grafting density improved the stability of the brush layer from 1 day to more than 7 days.
3 [23]	$P(PEGMA)_6$	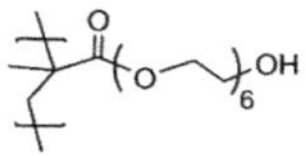	SiO_x	SI-ATRP	Deionized water, PBS, cell culture medium, 37 °C	The effect of the ATRP initiator structure on the detachment of $P(PEGMA)_6$, $P(PEGMEMA)_6$, PHEMA and PMAA brushes was studied. Incubation in MilliQ water at 37 °C showed that increasing the hydrophobicity of the initiator improves the stability of the polymer brushes. Introduction of a hydrophobic polymer block before initiating growth of the hydrophilic polymer block was found to increase the stability of polymer brushes.
	$P(PEGMEMA)_6$	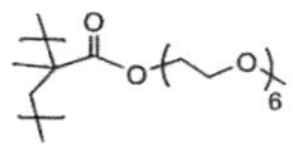				
	PHEMA	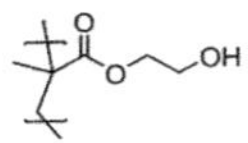				
	PEHMA-*b*-PMAA	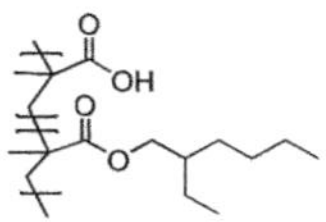				
	PMMA-*b*-PMAA	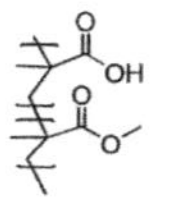				

(*continued*)

Table 6.1 (*continued*)

Entry	Polymer brush		Substrate	Polymerization method	Incubation conditions	Remarks
4[25]	PMMA-b-PSBMAm	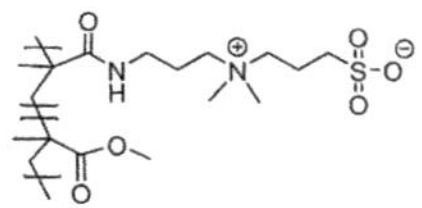	SiO_x	SI-ATRP	Seawater	PSBMAm brushes were prepared from SiO_x substrates by grafting from a protective hydrophobic PS or PMMA brush interface. PMMA-b-PSBMAm, PS-b-PSBMAm block copolymer brushes did not show any evidence for degrafting upon incubation in seawater for up to 3 months.
	PS-b-PSBMAm	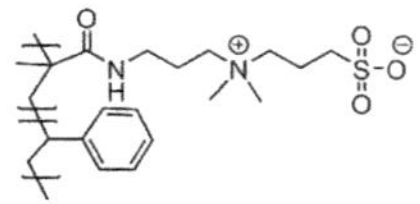				
	PSBMAm	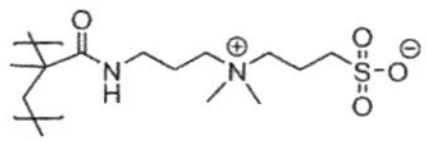				
5[26]	PDMAEMA	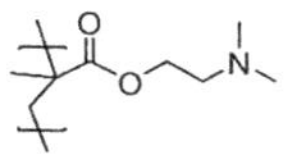	SiO_x	SI-ATRP	PBS with pH = 4, pH = 7.4, pH = 9	PDMAEMA and QPDMAEMA brushes were grown from both ester and non-ester based initiator modified SiO_x substrates. Incubation of the ester based initiator resulted in degrafting of polymer brushes especially at pH 9, whereas no detachment was observed for non-ester based initiator containing polymer brushes.
	QPDMAEMA	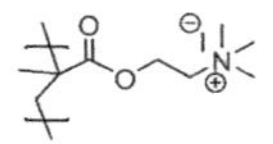		SI-FRP		
6[27]	PAA	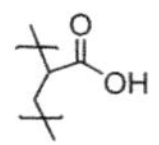	Mica	SI-ATRP	0.01 M NaCl solution at pH = 9	AFM analysis of PAA brushes grafted from mica revealed bare surfaces after 1 week when incubated in 0.01 M NaCl solution at pH 9.

7[28]	PAA-BODIPY	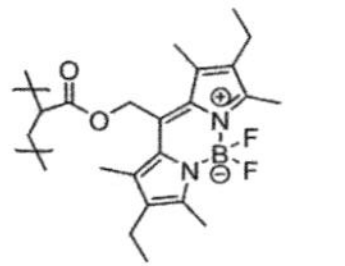	Glass	SI-ATRP	10 mM NaCl at pH = 6.5, pH = 7.5, pH = 8.5 pH = 9.5 pH = 10.5	Degrafting of BODIPY modified PAA brushes grown from glass substrates was studied in the absence and presence of 10 mM NaCl at different pH values. At pH 10 and with 10 mM NaCl, a decrease in the fluorescence intensity was observed indicative of degrafting.
8[29–32]	Carboxylated-PEGMA-*co*-PHEMA	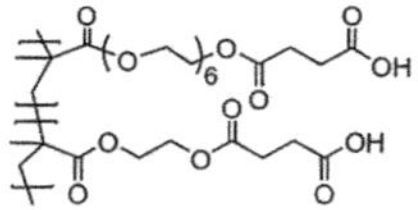	Au	SI-ARGET	PBS pH = 7.4, 25 °C	Effect of ionic strength, molecular composition and grafting density on the degrafting of carboxylated P(PEGMA-*co*-PHEMA) brushes grown from gold coated QCM sensors was studied. Wheras 54 nm and 75 nm thick brushes were stable in water, exposure of the 75 nm thick brush to PBS with 150 mM Na^+ resulted in degrafting. Increasing the HEMA content of brush from 52 to 75 mol% caused the polymer film layer to swell more and induced degrafting. Studies on the effect of grafting density did not show any relation between grafting density and degrafting.
9[34]	PS-SO_3H	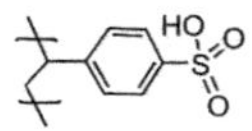	cPTFE ETFE PEEK	RIGP	Water 85 °C, 95 °C	PS-SO_3H polymer brushes were grafted from thermally stable polymer layers and incubated in water at 85 °C and 95 °C. UV and ^1H-NMR analysis of the detached polymer chains showed the same results with commercial authentic PSSA samples.

range of substrates. All of the examples of degrafting that have been reported so far involve hydrophilic polymer brushes that are exposed to aqueous media. The remainder of this section will discuss each of the examples listed in Table 6.1 in more detail.

Spontaneous degrafting of surface-tethered polymers was reported for the first time in 2007 by Deng and Zhu (Table 6.1, Entry 1).[21] These authors grafted hydrophilic, hyperbranched polyglycidol (HPG) *via* anionic ring-opening polymerization (AROP) from Au surfaces modified with binary mixtures of decanethiol, dodecanethiol, hexadecanethiol, 11-mercapto-1-undecanoic acid and 16-mercapto-1-hexadecanoic acid. The use of these binary thiol mixtures allowed changing of the local chemical environment and the interaction energy between neighboring HPG chains and the surface and to enhance, retard or prevent degrafting. Using tapping mode Atomic Force Microscopy (AFM) analysis, the authors observed systematic fluctuations in size and surface coverage of HPG molecules as a function of polymerization time, indicative of degrafting of HPG molecules. After 7.5 min. of growth time, surfaces functionalized with a binary mixture of 2.6% 11-mercapto-1-undecanoic acid and 97.4% decanethiol showed maximum thickness and coverage compared to shorter and longer growth time intervals. After 20 min. polymerization time, however, the same substrates revealed a bare Au surface. As the polymer molecules grew larger, they experienced more unfavorable interactions among neighboring HPG chains and with the hydrophobic surface. It was concluded that unfavorable polymer/surface and/or polymer/environment interactions provoked spontaneous degrafting of the surface-tethered HPG molecules during the polymerization reaction.

The first report on swelling-induced degrafting of polymer brushes prepared *via* surface-initiated atom transfer radical polymerization (SI-ATRP) from SiO_x substrates was published in 2008 by Klok *et al.* (Table 6.1, Entry 2).[22] Hydrophilic poly(poly(ethylene glycol) methacrylate) (PPEGMA) brushes were grown from trimethoxysilane-based ATRP initiator modified substrates and incubated both in pure water and cell culture medium at 37 °C. Upon incubation in water, no degrafting was observed. In contrast, incubation of the substrates in cell culture medium led to degrafting of the PPEGMA brushes from the substrates. Since the PPEGMA brushes are lightly cross-linked, degrafting resulted in wrinkling and delamination of the brush films. It was demonstrated that the grafting density had an influence on the degrafting process. Polymer brush films grown at lower grafting densities were more resistant to degrafting, presumably due to the fact that at lower grafting densities, the surface-tethered polymer chains were able to adopt a more collapsed conformation. Degrafting of polymer brushes was found to not only depend on grafting density, but also on the chemical structure of the anchoring point that tethered the polymers to the surface. In a subsequent study, the effect of the chemical structure of the ATRP initiator functionalized organosilane, which was used to modify the SiO_x substrates on the stability of hydrophilic brushes, was investigated (Table 6.1, Entry 3).[23] Three polymer brushes, namely poly(2-hydroxyethyl methacrylate) (PHEMA),

PPEGMA and poly(methacrylic acid) (PMAA) were grown from different ATRP initiator modified SiO_x substrates and incubated in cell culture medium until degrafting was observed. The ATRP initiators differed in the number of methylene groups that separated the anchoring group from the ATRP active group. Increasing the hydrophobicity of the ATRP initiator was found to decrease the rate of degrafting of the hydrophilic polymer brushes. The stability of the polymer brushes towards degrafting could be further enhanced by first growing a hydrophobic poly(methyl methacrylate) (PMMA) or poly(2-ethylhexyl methacrylate) (PEHMA) block from the ATRP initiator-modified surface before grafting the hydrophilic monomer. The hydrophobic block was proposed to prevent/retard access of water molecules to the brush–substrate interface, and hence prevent hydrolysis of siloxane, ester or amide bonds that were thought to be responsible for the degrafting of the polymer brushes. Based on the results of these two studies (Table 6.1, Entries 2 and 3), it was proposed that degrafting of hydrophilic polymer brushes involved hydrolytic bond cleavage at the brush/substrate interface (presumably the Si–O bond that linked the organosilane modified ATRP initiator to the silicon substrate and/or the amide bond that connected the ATRP initiating group with the aminopropyl trialkoxysilane anchoring moiety), which was facilitated by the solvent-induced stretching of the surface-grafted polymer chains (and hence a "mechanochemical(ly facilitated)" process) (Figure 6.2).[24]

Vancso and coworkers explored the use of a hydrophobic interlayer that covered the ATRP initiator modified substrate to improve the stability of zwitterionic brushes (Table 6.1, Entry 4).[25] By polymerizing sulfobetaine methacrylamide (SBMAm) from a polystyrene (PS) or poly(methyl methacrylate) (PMMA) brush substrate, hydrophilic brush layers were obtained, which were stable for more than 3 months in seawater.

In most of the initiators that are used in SI-ATRP, degrafting can potentially occur *via* cleavage of either the ester or amide bond that links the ATRP active group to the organosilane moiety or *via* cleavage of the siloxane bond that tethers the initiator to the surface. To eliminate the former, Genzer and coworkers prepared a non-ester based initiator (Scheme 6.1, Table 6.1, Entry 5).[26] Upon incubation in phosphate buffer solutions of pH 4, pH 7.4 and pH 9, poly(2-(dimethylamino)ethyl methacrylate) (PDMAEMA) and quaternized PDMAEMA (qPDMAEMA) brushes grafted from surfaces modified with the

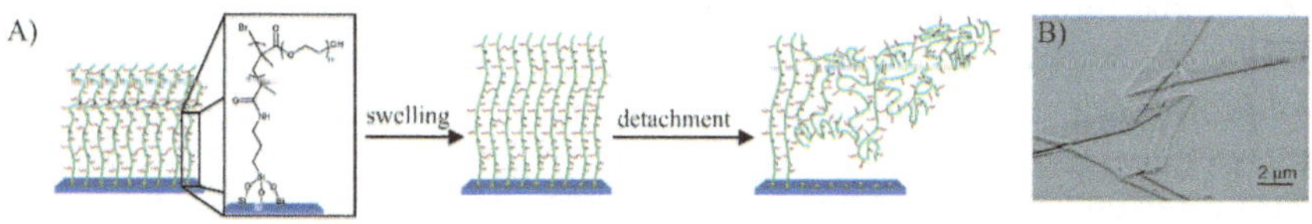

Figure 6.2 (A) Swelling-induced detachment of cross-linked PPEGMA brushes grafted from silicon substrates. (B) Scanning electron microscopy image of a densely grafted PPEGMA brush after 7 days of incubation in cell culture medium at 37 °C. Reprinted (adapted) with permission from ref. 24. Copyright 2015 American Chemical Society.

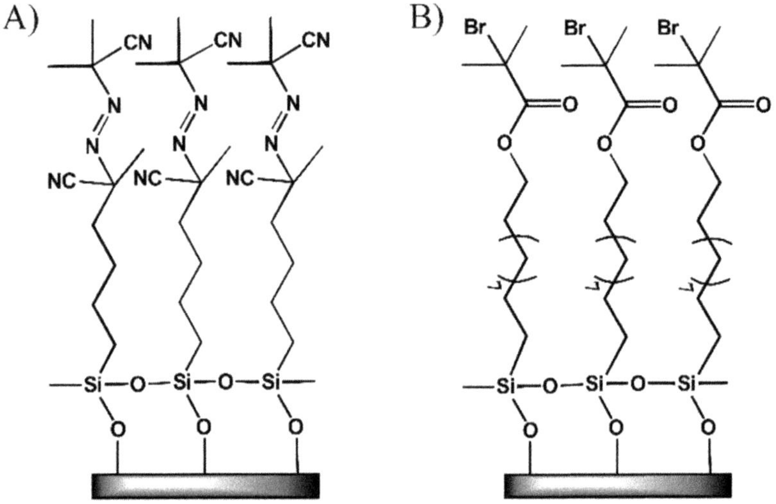

Scheme 6.1 Chemical structures of (A) non-ester-based radical initiator and (B) ester-based ATRP initiator by Genzer and coworkers.[26]

non-ester initiator did not reveal any significant variations in film thickness, whereas PDMAEMA and qPDMAEMA brushes grown from ester-containing ATRP initiator modified surfaces did degraft, with the extent of degrafting being dependent on the pH of the aqueous medium. This further supports the hypothesis (*vide supra*) that ester bond hydrolysis and the brush substrate interfaces are involved in these grafting processes.

Another weak polyelectrolyte brush that has been investigated is poly(acrylic acid) (PAA). Giasson and coworkers studied degrafting of PAA brushes grown from mica and glass surfaces at higher pH (Table 6.1, Entries 6 and 7).[27,28] Whereas a bare mica surface was observed by AFM after incubating PAA brush modified surfaces for one week in 10 mM NaCl solution at pH 9, there was no observable cleavage from the substrates incubated at pH 9 in the absence of NaCl salt. This study confirmed the enhancing effect of ionic strength on the swelling and the detachment of polyelectrolyte brushes. In a second study by the same authors, degrafting of PAA brushes was monitored in real time from glass substrates (Table 6.1, Entry 7).[28] For this purpose some of the PAA repeating units were modified with a boron-dipyrromethene (BODIPY) derivative. Monitoring the intensity of the BODIPY emission at 550 nm while immersing the substrates in solutions covering a range of pH values and ionic strengths was used to follow the degrafting of polymer chains *via* total internal reflection fluorescence (TIRF) microscopy. In the absence of salt, PAA brushes did not exhibit any observable decrease in the fluorescence intensity at pH values ranging from 6.5 to 10.5, but addition of 10 mM NaCl revealed a significant decrease in the fluorescence intensity indicating the degrafting of polymer chains. Moreover, AFM thickness measurements conducted before and after degrafting supported the TIRF microscopy results and indicated an 80% reduction in film thickness. The combination of high pH and salt caused a change in the conformation of the polymer brush layer to a swollen state while creating osmotic pressure at the polymer substrate interface.

Ma *et al.* investigated the swelling-induced degrafting of Au–S tethered, carboxylated poly(oligo(ethylene glycol)methacrylate-*co*-2-hydroxyethyl methacrylate) (P(PEGMA-*co*-HEMA)) brushes using quartz crystal microbalance (QCM) experiments (Table 6.1, Entry 8).[29] Neutral P(PEGMA-*co*-HEMA) brushes prepared from a monomer feed containing equimolar amounts of PEGMA and HEMA were grafted on QCM sensors *via* surface-initiated activators regenerated by electron transfer atom transfer radical polymerization (SI-ARGET ATRP) and subsequently converted to weak polyelectrolytes by reacting with succinic anhydride. During this study two samples were used with thicknesses of 54 and 75 nm. While exposing the carboxylated P(PEGMA-*co*-HEMA) brush films to MilliQ water (pH = 6.2) resulted in a decrease in the frequency indicating that the polymer brushes absorb more water, phosphate buffer saline (PBS, pH = 7.4, $[Na^+]$ = 150 mM) caused a sudden increase in the frequency for the 75 nm thick polymer brush layer, suggesting instantaneous degrafting of polyelectrolyte chains from the Au surface. In contrast, exposing a 54 nm thick P(PEGMA-*co*-HEMA) brush to PBS did not result in degrafting, but rather in a further frequency decrease reflecting a further swelling of these films. By evaluating a series of samples with different initial film thicknesses, the authors identified 75 nm as the critical thickness for degrafting to occur.

In subsequent studies, the same group investigated the effects of ionic strength, molecular composition and grafting density on the effective critical thickness for degrafting of carboxylated P(PEGMA-*co*-HEMA) brushes (Table 6.1, Entry 8).[30–32] Increasing the $[Na^+]$ concentration in PBS from 0.075 mM to 7.5 mM during the QCM experiments on brushes with dry thicknesses more than 75 nm resulted in a decrease in frequency due to uptake of ions in the brush layer without, however, inducing degrafting. In contrast, degrafting was observed at $[Na^+]$ concentrations higher than 15 mM. The authors found that the effective critical wet thickness for degrafting to occur (T_{eff}) could be described by the following equation $T_{eff} = (0.32\ln([Na^+]) + 2.53) \times T_{COOH,dry}$, where $T_{COOH,dry}$ is the thickness of the polymer brush film as measured by ellipsometry in air. In addition to Au, the authors also grafted carboxylated P(PEGMA-*co*-HEMA) brushes from Pt and Ag substrates. It was found that $T_{COOH,dry}$ decreased from >100 nm on Pt to 75 nm on Au to 47 nm on Ag, which was consistent with the relative energies that were calculated for the absorption of alkanethiols on these different surfaces.

The effect of the chemical composition of the P(PEGMA-*co*-HEMA) brushes on degrafting was studied with two samples with dry film thicknesses of 57 nm and 75 nm and containing 75, respectively 52 mol% HEMA. The predicted swelling ratios of these samples were estimated to 4.10 for the 75 nm film containing 52% HEMA and to 4.85 for the 57 nm thick film composed of 75% HEMA. The $T_{COOH,PBS,critical}$ of the two samples were predicted to 236 nm

(75 mol% HEMA) and 249 nm (52 mol% HEMA). The fact that the $T_{COOH,PBS,critical}$ values were relatively close in spite of the difference in dry film thicknesses was interpreted as evidence for the importance of swelling (which increased with increasing HEMA content) on the degrafting process.

The effect of grafting density on the stability of carboxylated P(PEGMA-*co*-HEMA) brushes grown from Au substrates was examined in the following study of the same group.[32] Polymer films were prepared from substrates that were modified with mixtures of active and dummy initiators to obtain surfaces with 2.5, 5, 7.5, 10 and 100% grafting densities. QCM experiments carried out in PBS (pH = 7.4, $[Na^+]$ = 150 mM) revealed degrafting at all five grafting densities. While $T_{critical,dry}$ for the onset of degrafting was found to increase with increasing grafting density, the changes were relatively modest. This is in contrast to previous studies on PPEGMA brushes grafted from SiO_x substrates where variation in grafting density was found to have a pronounced effect on degrafting.[22] The apparent discrepancy between the effects of grafting density on the degrafting of polymer brushes grown from gold surfaces as compared to SiO_x may be related to the difference in bond dissociation energies between the Au–S (253.6 ± 14.6 kJ mol^{-1}) and the Si–O bond (799.6 ± 13.4 kJ mol^{-1}).[33]

Enomoto *et al.* investigated the degrafting of hydrophilic brushes grown from hydrophobic polymer substrates, *viz.* cross-linked poly(tetrafluoroethylene) (cPTFE), poly(ethylene-*co*-tetrafluoroethylene) (ETFE) and poly(ether ether ketone) (PEEK) (Table 6.1, Entry 9).[34] Poly(styrenesulfonic acid) (PSSA) brushes were grown from these substrates *via* radiation-induced graft polymerization (RIGP). Degrafting of these brushes was studied by incubating samples in water at 85 °C. The detached PSSA extracts obtained after incubation times of 100, 200, 300 and 500 h were analyzed *via* UV-Visible (UV-Vis) spectroscopy. The UV-Vis spectra of these samples showed the same profile as the spectrum that was obtained for an authentic PSSA sample. NMR analysis of the degrafted material revealed the typical PSSA signals, from which it was concluded that the grafts were detached as PSSA structure from the polymer substrates. From these analyses, the authors concluded that the observed degrafting involved scission of C–C bonds at the interface between the polymer substrate and the PSSA grafts.

6.3 Summary and Conclusions

Polymer brushes grafted by surface-initiated controlled radical polymerization (SI-CRP) represent a versatile platform to engineer surface and interface properties. It is well established that several of the attractive material properties of densely grafted, hydrophilic polymer brushes, such as their low friction coefficients and ability to prevent biofouling are related to the stretched conformation of the surface-tethered polymer chains. What is less well known is that the stretched chain conformation, especially upon swelling in a good solvent, can also alter the chemical reactivity of these surface-grafted polymers. This chapter summarized results from a number of studies that have reported degrafting of polymer brush films, putatively due to swelling-induced, mechanochemically facilitated cleavage reactions at the polymer brush/substrate interface. These results are of interest for various reasons. First, the realization of, and a first understanding of, these degrafting reactions paves the way towards robust polymer brush films. Second, the observation of these degrafting processes underlines the potential of SI-CRP as an alternative tool to mechanochemically activated polymers, without the need for external force fields such as flow or ultrasound. Thus, this represents an expansion of the current polymer mechanochemistry toolbox and may provide new opportunities to study the chemistry and reactivity of synthetic polymers upon exposure to mechanical forces.

References

1. V. N. Fery, R. Laible and K. Hamann, *Angew. Makromol. Chem.*, 1973, **34**, 81.
2. T. Otsu, T. Ogawa and T. Yamamoto, *Macromolecules*, 1986, **19**, 2087.
3. O. Azzaroni, *J. Polym. Sci., Part A: Polym. Chem.*, 2012, **50**, 3225.
4. N. Ayres, *Polym. Chem.*, 2010, **1**, 769.
5. L. Moroni, M. Klein Gunnewiek and E. M. Benetti, *Acta Biomater.*, 2014, **10**, 2367.
6. C. J. Galvin and J. Genzer, *Prog. Polym. Sci.*, 2012, **37**, 871.
7. S. T. Milner, *Science*, 1991, **251**, 905.
8. B. Zhao and W. J. Brittain, *Prog. Polym. Sci.*, 2000, **25**, 677.
9. J. Pyun, T. Kowalewski and K. Matyjaszewski, *Macromol. Rapid Commun.*, 2003, **24**, 1043.
10. W. J. Brittain and S. Minko, *J. Polym. Sci., Part A: Polym. Chem.*, 2007, **45**, 3505.
11. R. Barbey, L. Lavanant, D. Paripovic, N. Schüwer, C. Sugnaux, S. Tugulu and H.-A. Klok, *Chem. Rev.*, 2009, **109**, 5437.
12. M. E. Welch and C. K. Ober, *J. Polym. Sci., Part B: Polym. Phys.*, 2013, **51**, 1457.
13. M. Chen, W. H. Briscoe, S. P. Armes, H. Cohen and J. Klein, *ChemPhysChem*, 2007, **8**, 1303.
14. M. Chen, W. H. Briscoe, S. P. Armes and J. Klein, *Science*, 2009, **323**, 1698.
15. I. E. Dunlop, W. H. Briscoe, S. Titmuss, R. M. J. Jacobs, V. L. Osborne, S. Edmondson, W. T. S. Huck and J. Klein, *J. Phys. Chem. B*, 2009, **113**, 3947.
16. A. Hucknall, S. Rangarajan and A. Chilkoti, *Adv. Mater.*, 2009, **21**, 2441.
17. J. E. Raynor, J. R. Capadona, D. M. Collard, T. A. Petrie and A. J. García, *Biointerphases*, 2009, **4**, FA3–FA16.
18. I. Banerjee, R. C. Pangule and R. S. Kane, *Adv. Mater.*, 2011, **23**, 690.
19. C. Blaszykowski, S. Sheikh and M. Thompson, *Chem. Soc. Rev.*, 2012, **41**, 5599.
20. J. Rühe, in *Polymer Brushes*, ed. R. C. Advincula, W. J. Brittain, K. C. Caster and J. Rühe, Wiley-VCH Verlag GmbH & Co. KGaA, Weinheim, 2004, vol. 1, pp. 1–31.

21. Y. Deng and X. Y. Zhu, *J. Am. Chem. Soc.*, 2007, **129**, 7557.
22. S. Tugulu and H.-A. Klok, *Biomacromolecules*, 2008, **9**, 906.
23. D. Paripovic and H.-A. Klok, *Macromol. Chem. Phys.*, 2011, **212**, 950.
24. H.-A. Klok and J. Genzer, *ACS Macro Lett.*, 2015, **4**, 636.
25. R. Quintana, M. Gosa, D. Jańczewski, E. Kutnyanszky and G. J. Vancso, *Langmuir*, 2013, **29**, 10859.
26. E. D. Bain, K. Dawes, A. E. Özçam, X. Hu, C. B. Gorman, J. Šrogl and J. Genzer, *Macromolecules*, 2012, **45**, 3802.
27. B. Lego, W. G. Skene and S. Giasson, *Macromolecules*, 2010, **43**, 4384.
28. O. Borozenko, R. Godin, K. L. Lau, W. Mah, G. Cosa, W. G. Skene and S. Giasson, *Macromolecules*, 2011, **44**, 8177.
29. Y. Zhang, J. He, Y. Zhu, H. Chen and H. Ma, *Chem. Commun.*, 2011, **47**, 1190.
30. Y. Zhang, B. Lv, Z. Lu, J. He, S. Zhang, H. Chen and H. Ma, *Soft Matter*, 2011, **7**, 11496.
31. Y. Zhu, B. Lv, P. Zhang and H. Ma, *Chem. Commun.*, 2011, **47**, 9855.
32. B. Lv, Y. Zhou, W. Cha, Y. Wu, J. Hu, L. Li, L. Chi and H. Ma, *ACS Appl. Mater. Interfaces*, 2014, **6**, 8313.
33. Y. R. Luo, *Handbook of Bond Dissociation Energies in Organic Compounds*, CRC Press, Florida, 2003, pp. 65–70.
34. K. Enomoto, S. Takahashi, T. Iwase, T. Yamashita and Y. Maekawa, *J. Mater. Chem.*, 2011, **21**, 9343.

CHAPTER 7

Coupling Mechanics to Chemical Reactions to Create "Materials that Compute"

VICTOR V. YASHIN[a], STEVEN P. LEVITAN[b] AND ANNA C. BALAZS*[a]

[a]Department of Chemical Engineering, University of Pittsburgh, Pittsburgh, PA 15261, USA; [b]Department of Electrical and Computer Engineering, University of Pittsburgh, Pittsburgh, PA 15261, USA
*E-mail: balazs@pitt.edu

7.1 Introduction

The development of stimuli-responsive materials has enabled the fabrication of synthetic systems that can sense and respond to changes in the environment in ways that mimic the behavior of biological organisms and, hence, has facilitated the creation of novel biomimetic sensors and actuators.[1] Researchers are now attempting to design the next generation of adaptive materials that encompass even greater functionality. In particular, materials of the future will integrate sensing, actuation, computation and communication.[2] Current realizations of these multi-functional materials utilize a combination of embedded sensors and microelectronics to perform these tasks[2] and, thus, require external power sources for their operation. For these systems to mimic the living organism that inspired their design, they should, however, operate in a

Polymer Chemistry Series No. 26
Mechanochemistry in Materials
Edited by Yoan C. Simon and Stephen L. Craig

Published by the Royal Society of Chemistry, www.rsc.org

relatively autonomous way. The development of self-powered materials that can autonomously sense, actuate, communicate and compute could ultimately lead to new forms of smart materials such as clothing, prosthetics and robotic systems.

The challenge in designing such advanced systems is isolating materials whose inherent properties enable autonomous functionality. Here, we describe our recent efforts to develop theoretical models for the design of "materials that compute", which perform sensing, actuation, communication and computation in a self-sustained and autonomous manner. Namely, these materials systems could perform these tasks on the scale of hours without the need for an external power supply.

Our computing materials system integrates the distinctive properties of self-oscillating polymer gels and piezoelectric layers. We specifically consider polymer gels undergoing the oscillatory Belousov–Zhabotinsky (BZ) reaction.[3] Notably, these oscillating BZ gels are powered by an internal chemical reaction and, hence, do not need extensive peripheral equipment to operate. This autonomous, self-oscillatory response happens thanks to a ruthenium catalyst, covalently bonded to the polymers. The oscillating BZ reaction induces a periodic oxidation and reduction of the anchored metal ion; the hydrating effect of the oxidized metal catalyst triggers the gel expansion, followed by a contraction when the catalyst is in the reduced state. The chemical energy emanating from the BZ reaction sustains the mechanical oscillations of these active, chemo-responsive gels. The gels pulsate in a self-sustained manner until the reagents in the host solution are consumed, but the system can be "re-fueled" by simply replenishing these solutes. Furthermore, the BZ gels not only undergo chemo-mechanical transduction, but also mechano-chemical transduction, where mechanical force can resuscitate or alter the chemical pulsations. The BZ gels are also photosensitive, so that light can be used to tailor the dynamic behavior of the materials.

We also take advantage of two distinctive properties of piezoelectric electric materials. Namely, these materials produce a voltage when they are deflected and conversely, the materials undergo a deflection in response to an applied voltage. In this chapter, we describe how the BZ gel and a piezoelectric (PZ) cantilever can be combined to create our fundamental BZ-PZ unit. Within each unit, the oscillating BZ gels caused the overlaying PZ layer to deflect and, thereby, generate a voltage. When this unit was connected in a circuit to another BZ-PZ unit, the generated voltage caused the PZ cantilever in the second unit to deflect and perturb the oscillations of the underlying BZ gel. In this manner, the units sense, actuate and communicate with each other. After multiple interactions, the oscillations of these BZ-PZ units become synchronized, as we discuss further below. It is this synchronization of oscillators that allows us to perform computation.

It is important to emphasize that these BZ-PZ "materials that compute" do not perform Boolean operations associated with traditional computers. Rather, they operate by performing pattern-matching operations. In particular, we will ultimately harness the synchronization of the oscillations of the BZ-PZ units to achieve associative processing between assortments of stored patterns and input stimuli such as light, pressure or chemistry. The effective resonance between the input and the BZ-PZ oscillators will be utilized for our basic computation paradigm.

Below, we briefly summarize our approach for simulating BZ gels. Before describing our theoretical and computational models for the BZ-PZ systems, we first develop and apply our formalism to a simpler system where an external periodic force is applied to a single BZ gel. This system has been examined experimentally and, hence, we can validate our model by comparing our findings to the experimental results. As shown below, there is excellent agreement between our calculations and the experimental observations. This agreement lends credence to our predictions on the behavior of the BZ-PZ systems, which have not yet been fabricated. Hence, our predictions can provide valuable guidelines for creating these computing platforms. Recent studies on the behavior of a humidity-responsive gel on piezoelectric substrates[4] and the fabrication of wearable piezoelectric patches that can measure small degrees of strain[5,6] indicate that the systems described herein could be fabricated in the near future.

7.2 Modeling Self-oscillating Gels

We consider a BZ gel sample that is immersed in a solution of BZ reactants under conditions that produce autonomous chemo-mechanical oscillations of the gel. Simultaneously, a time-dependent compressive force is applied to the self-oscillating sample (see Figure 7.1a). We assume that the gel is sufficiently small that the chemo-mechanical oscillations are spatially uniform. To capture the gel dynamics, the equations describing the chemical kinetics of the BZ reaction are coupled to the equations characterizing the variations in the size of the gel due to the inherent reaction and external force. Below, we first discuss the governing equations that describe the behavior of gels undergoing the BZ reaction. We then detail the expressions for capturing the response of this self-oscillating gel to an applied compression.

7.2.1 Kinetics of the BZ Reaction in a Polymer Gel

We assume that the gel is formed from poly(*N*-isopropylacrylamide) (PNIPAAm) chains containing grafted ruthenium metal-ion catalysts.[3] The kinetics of the BZ reaction is described by a modification of the Oregonator model,[7] formulated in terms of the dimensionless concentrations of the key reaction intermediate u ($HBrO_2$, the activator), and the oxidized metal-ion catalyst v (Ru^{3+} in the case considered here). The modified model[8,9] accounts for the dependence of the BZ reaction rates on the volume fraction of polymer ϕ, and on the total concentration of catalyst grafted to the network.

(a)

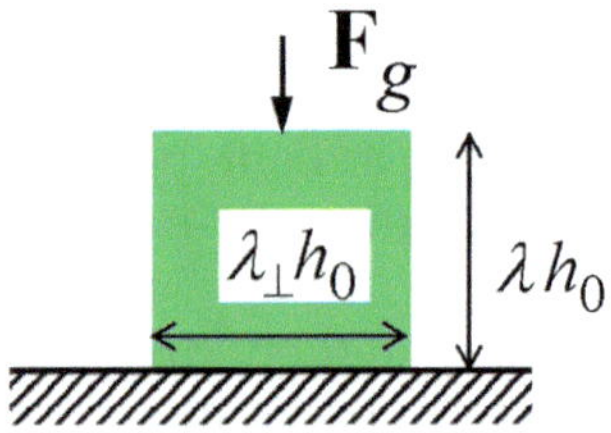

(b)

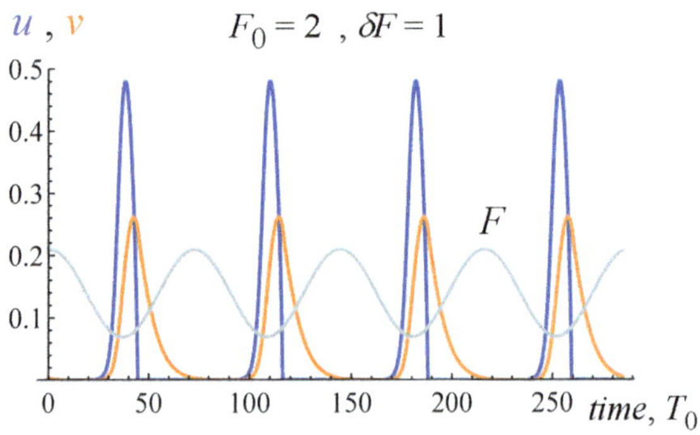

(c)

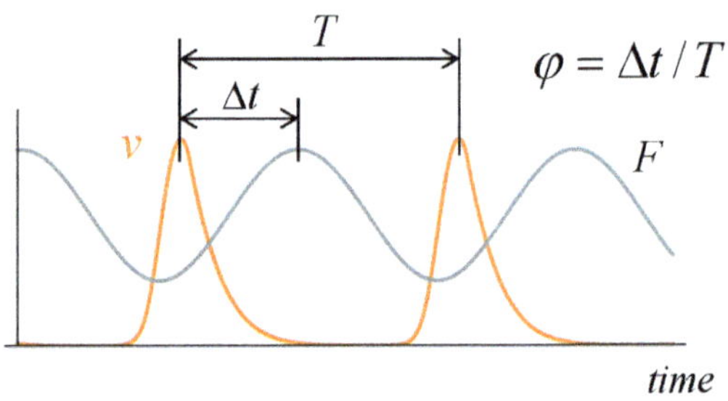

Figure 7.1 Force-controlled entrainment of self-oscillating BZ gels. (a) Schematics of gel swelling under the applied compressive force $\mathbf{F}_g$. The longitudinal and transverse degrees of swelling (strains) are denoted by λ and $\lambda_\perp$, respectively. (b) Oscillations of the BZ reactants u (blue) and v (orange) entrained to the oscillations of force F (gray). The sinusoidal force is shown using an arbitrary scale. (c) Definition of the phase difference φ between the BZ oscillations and the periodic force.

In small gel samples, diffusion of the dissolved activator throughout the gel occurs faster than variations of u and v due to the BZ reaction; hence, the reaction kinetics is given by:[10,11]

$$(1-\phi)\mathrm{d}/\mathrm{d}t[u(1-\phi)^{-1}] = F_{\mathrm{BZ}}(u,v,\phi) \tag{7.1}$$

$$\phi\,\mathrm{d}/\mathrm{d}t(v\phi^{-1}) = G_{\mathrm{BZ}}(u,v,\phi) \tag{7.2}$$

The reaction rates F_{BZ} and G_{BZ} in eqn (7.1) and (7.2) are determined according to the modified Oregonator model[8,9] as follows:

$$F_{BZ}(u,v,\phi) = (1-\phi)^2 wr - u^2 - (1-\phi) f v \frac{u - q(1-\phi)^2}{u + q(1-\phi)^2} \tag{7.3}$$

$$G_{BZ}(u,v,\phi) = \varepsilon_{BZ}[(1-\phi)^2 wr - (1-\phi)v] \tag{7.4}$$

The above reaction rates depend on the dimensionless concentrations of the reduced catalyst Ru^{2+} and the radical BrO_2*, which are denoted by r and w, respectively.[9] The concentration r is calculated as: $r = c_{Ru}\phi_0^{-1}\phi - v$, where c_{Ru} and ϕ_0 are the catalyst concentration and volume fraction of polymer in the undeformed gel, respectively. The value of w, the concentration of the radical, is found as follows: $w = \mu_{BZ}(r^2 + 2u/\mu_{BZ})^{1/2} - \mu_{BZ}r$, where μ_{BZ} is a dimensionless parameter (see ref. 9 for further details). Finally, the stoichiometric factor f and the dimensionless parameters q and ε_{BZ} have the same meaning as in the original Oregonator model.[7]

7.2.2 Gel Swelling in the Presence of an External Force

The internal stresses acting on the swollen polymer network are described by the following general stress–strain equation:[10,11]

$$\hat{\boldsymbol{\sigma}} = -\pi_{osm}\hat{\mathbf{I}} + \hat{\boldsymbol{\sigma}}_{el}, \tag{7.5}$$

where π_{osm} is the osmotic pressure of the polymer, $\hat{\mathbf{I}}$ is the unity tensor and $\hat{\boldsymbol{\sigma}}_{el}$ is the contribution to the stress tensor from the elasticity of the cross-linked polymer network. The osmotic pressure of the polymer is calculated as:

$$\pi_{osm}(\phi,v) = \pi_{FH}(\phi) + \chi^* v\phi \tag{7.6}$$

Here, $\pi_{FH}(\phi) = -[\phi + \log(1-\phi) + \chi(\phi)\phi^2]$ arises from the Flory–Huggins theory, where $\chi(\phi)$ describes the polymer–solvent interactions, and the term $\chi^* v\phi$ describes the hydrating effect of the oxidized catalyst on the responsive gel. The strength of the hydrating effect is controlled by the coupling parameter $\chi^* > 0$. The gel elasticity contribution to the stress tensor is calculated using the Flory model of rubber elasticity to obtain:[11]

$$\hat{\boldsymbol{\sigma}}_{el} = -c_0\phi(2\phi_0)^{-1}\hat{\mathbf{I}} + c_0\phi_0^{-1}\phi\hat{\mathbf{B}} \tag{7.7}$$

Here, c_0 is the cross-link density, ϕ_0 and ϕ are the volume fractions of polymer in the undeformed and deformed gel, respectively, and $\hat{\mathbf{B}}$ is the Finger strain tensor.[12] In the absence of an external force, the equilibrium swelling is described by the force balance equation: $\hat{\boldsymbol{\sigma}} = 0$.

In the case where an external force $\mathbf{F}_g$ is acting on the gel as shown in Figure 7.1, the gel deformation is described by the strain tensor

$$\hat{\mathbf{B}} = \begin{pmatrix} \lambda^2 & 0 & 0 \\ 0 & \lambda_\perp^2 & 0 \\ 0 & 0 & \lambda_\perp^2 \end{pmatrix},$$

where λ and $\lambda_\perp$ are the degrees of swelling in the longitudinal and transverse directions, respectively. The equilibrium degrees of swelling are determined by balancing all the forces in the two directions, namely,

$$c_0\phi_0^{-1}\phi(\lambda^2 - 1/2) + (h_0\lambda_\perp)^{-2}F_g = \pi_{osm}(\phi,\nu) \tag{7.8}$$

$$c_0\phi_0^{-1}\phi(\lambda_\perp^2 - 1/2) = \pi_{osm}(\phi,\nu) \tag{7.9}$$

Here, h_0 is the undeformed size of the gel. The osmotic pressure π_{osm} is calculated according to eqn (7.6). Finally, the volume fraction of polymer depends on the degrees of swelling λ and $\lambda_\perp$ as $\phi = \phi_0\lambda^{-1}\lambda_\perp^{-2}$. Using eqn (7.8) and (7.9), we can determine the values of λ and $\lambda_\perp$ as functions of the concentration of oxidized catalyst ν and the force F_g. It is convenient to re-write eqn (7.8) and (7.9) in the following form:

$$\phi = \phi_0\lambda^{-2}(\lambda + F)^{-1} \tag{7.10}$$

$$c_0\lambda^{-1} = \pi_{osm}(\phi,\nu) + c_0\phi(2\phi_0)^{-1}, \tag{7.11}$$

where $F = (c_0h_0^2)^{-1}F_g$ is the dimensionless force, and solve the above equations for ϕ and λ.

7.2.3 Values of Parameters Used in the Calculations

We take the chemical composition of the BZ substrate, the volume fraction of polymer and catalyst concentration in the undeformed gel to be the same as used in the experiments described in ref. 8, so the corresponding dimensionless parameters are estimated to be $\varepsilon_{BZ} = 0.212$, $q = 9.52 \times 10^{-5}$, $\mu_{BZ} \approx 7 \times 10^2$, $\phi_0 = 0.16$ and $c_{Ru} = 0.855$. The stoichiometric parameter f is an adjustable parameter in the model and, here, we take $f = 1$.

The cross-link density in the undeformed BZ gel is the same as in ref. 8, namely, $c_0 = 3 \times 10^{-3}$. The polymer–solvent interactions are characterized by the function $\chi(\phi) = 0.338 + 0.518\phi$, which is known to describe the PNIPAAm-water interaction at 20 °C.[13] The interaction parameter χ^*, which accounts for the hydrating effect of the oxidized catalyst, is an adjustable parameter of the model; we set $\chi^* = 0.105$ in this study. The same value of χ^* was used in our previous publications.[9,11] Finally, the undeformed gel size is $h_0 = 0.5$mm; this choice guarantees the spatial uniformity of chemo-mechanical oscillations.[14]

7.3 Modeling Force-controlled Entrainment of BZ Gels

We examine how the dynamic behavior of a BZ gel becomes entrained to periodic deformations of the sample by a compressive load $F(t)$ (*i.e.* by force-controlled deformations). This mechanical action induces periodic variations in the polymer volume fraction ϕ that, in turn, affects the ϕ-dependent BZ reaction kinetics (see eqn (7.1) and (7.2)) and can result in entraining the BZ gel dynamics. The strain-controlled entrainment of BZ gels is discussed in ref. 15.

Figure 7.1b shows the BZ oscillations in u and v within a gel sample that has become entrained by the force oscillations. The effect of the periodic mechanical action on the BZ gel dynamics is characterized by the phase difference φ between the BZ oscillations and cycles of the mechanical action. The phase difference is determined as schematically shown in Figure 7.1c. The initial value of φ is arbitrary within the interval $0 \le \varphi < 1$. The external mechanical action affects the BZ gel dynamics so φ changes in time. If entrainment is attainable, φ evolves towards some stationary value, which characterizes the state of entrainment. The frequency of oscillation of the BZ gel with no external periodic stimulation and the frequency of the external action do not necessarily coincide. Establishing a stationary phase difference signifies entrainment because in this case the BZ oscillations occur with the frequency imposed by the external action.

Figure 7.1b reveals a salient feature of an entrained sample: the oxidation stage of the BZ reaction cycle takes place when the gel is less compressed as the concentration of oxidized catalyst v exhibits maxima close to the minima in force F (Figure 7.1b). Figure 7.1b is obtained by solving eqn (7.1) and (7.2) simultaneously with eqn (7.10) and (7.11), where the force is $F(t) = F_0 + \delta F \sin(2\pi\nu t)$ at $F_0 = 2$ and $\delta F = 1$. The force oscillation frequency ν is taken equal to that of the BZ gel oscillation at the given static force, but without the periodic stimulation, *i.e.* at $F_0 = 2$ and $\delta F = 0$.

Evolution of the entrainment process, *i.e.* the evolution of the phase difference φ, can be modeled directly by solving the system of equations for the BZ reaction kinetics, eqn (7.1) and (7.2), and for the gel size, eqn (7.10) and (7.11), followed by calculation of $\Delta\varphi$ according to the definition shown in Figure 7.1c. If the external mechanical action is sufficiently weak, then we can employ the phase dynamics approach.[16–18] Below, we utilize this general approach to derive specific phase dynamics equations for BZ gels that are subjected to relatively weak mechanical deformation.

7.3.1 Phase Dynamics Equations for Mechanically Deformed BZ Gels

The phase dynamics approach is an approximation used to describe the behavior of an oscillator forced by a periodic input in the limit of weak input.[16–18] The system is assumed to be governed by the following equation

$$\mathrm{d}\mathbf{x}/\mathrm{d}t = \mathbf{f}(\mathbf{x}) + \mu\mathbf{g}(\mathbf{x})p(t). \tag{7.12}$$

Here, the vector $\mathbf{x}$ is the set of variables describing the oscillator (in our case, the concentrations of the activator, u, and oxidized catalyst, v), $p(t)$ is the external time-periodic input and the parameter μ controls the input strength. The second term on the right-hand side of eqn (7.12) represents the linear response of the system to the periodic input $p(t)$, which is assumed to be weak, and the magnitude of the response depends on the system state through the vector function $\mathbf{g}(\mathbf{x})$. In the case of no forcing ($p = 0$), the oscillator exhibits the T-periodic stable limit cycle behavior $\bar{\mathbf{x}}(t + T) = \bar{\mathbf{x}}(t)$. If the input is weak ($\mu \ll 1$), the forcing results primarily in a deviation of the phase of oscillation, and the solution of eqn (7.12) takes the form $\mathbf{x}(t) = \bar{\mathbf{x}}(t + \varphi)$, where φ is the oscillation phase.

The temporal evolution of the phase is described by the following equation:[18]

$$\mathrm{d}\phi/\mathrm{d}t = \mu\mathbf{Q}(t + \varphi)\cdot\mathbf{g}(\bar{\mathbf{x}}(t + \varphi))p(t) \tag{7.13}$$

The T-periodic vector function $\mathbf{Q}(\theta)$ is the phase response curve (PRC),[18] which is a measure of the sensitivity of the system to external stimulation throughout its oscillation cycle. Specifically, the PRC characterizes the magnitude of the phase shift, which is induced by the stimulus and depends on the exact timing of the stimulation relative to the phase of oscillation.[18] The PRC can be determined using Malkin's method,[17,18] where $\mathbf{Q}(\theta)$ is the solution of the "adjoint" equation

$$\mathrm{d}\mathbf{Q}/\mathrm{d}t = -\{D\mathbf{f}(\bar{\mathbf{x}}(t))\}^{\mathrm{T}}\mathbf{Q}, \tag{7.14}$$

under the initial condition $\mathbf{Q}(0)\cdot\mathbf{f}(\bar{\mathbf{x}}(0)) = 1$. In eqn (7.14), $\{D\mathbf{f}(\bar{\mathbf{x}}(t))\}$ is the Jacobian matrix for a free-running oscillator (see eqn (7.12) at $p = 0$) calculated along the limit cycle.

Solutions of eqn (7.13) exhibit both fast dynamics on the time scale of the period of oscillation T and slow dynamics on the time scale of μ^{-1}. The entraining phenomena take place on the slow time scale. To extract the slow dynamics, time averaging is applied to eqn (7.13). If the period of oscillation of the input, T_{p}, is equal to that of the free-running oscillator, T, then the averaging procedure yields the following result:[17,18]

$$d\varphi/dt = \mu H(\varphi), \tag{7.15}$$

where

$$H(\varphi) = T^{-1}\int_0^T \mathbf{Q}(t)\cdot\mathbf{g}(\overline{\mathbf{x}}(t))p(t-\varphi)dt \tag{7.16}$$

If the periods of oscillation of the input and free-running oscillator are slightly different, the time-averaged phase dynamics is described by the following equation:[17]

$$d\varphi/dt = \Delta v + \mu H(\varphi), \tag{7.17}$$

where $\Delta v = T^{-1} - T_p^{-1}$ is the detuning, and $H(\varphi)$ is given by eqn (7.16). According to eqn (7.16), the phase φ is calculated relative to that of the external periodic input, so the value of φ is in effect the phase difference between the BZ gel oscillations and the external forcing.

To apply the above formalism to the BZ gel under an external action, we first represent the equations for the BZ gel dynamics in the form of eqn (7.12). We introduce the partial variables for the BZ reactants $\mathrm{u} = u(1-\phi)^{-1}$ and $\mathrm{v} = v\phi^{-1}$, and write eqn (7.1) and (7.2) for the BZ reaction kinetics as

$$\frac{d}{dt}\begin{pmatrix}\mathrm{u}\\ \mathrm{v}\end{pmatrix} = \begin{pmatrix}\mathrm{F}(\mathrm{u},\mathrm{v},\phi)\\ \mathrm{G}(\mathrm{u},\mathrm{v},\phi)\end{pmatrix} \tag{7.18}$$

where $\mathrm{F} = (1-\phi)^{-1}F_{BZ}$ and $\mathrm{G} = \phi^{-1}G_{BZ}$. This assumes that the diffusion of the dissolved activator, u, throughout the gel occurs faster than variations of u and v due to the BZ reaction. With no oscillatory action imposed on the gel, eqn (7.18) is solved in conjunction with eqn (7.10) and (7.11) to yield the volume fraction of polymer as a function of the partial concentration of oxidized catalyst, $\phi(\mathrm{v})$. The external mechanical action perturbs the volume fraction of polymer in the gel, $\phi(\mathrm{v}) + \delta\phi$. If the perturbation $\delta\phi$ is small, eqn (7.18) can be expanded into a Taylor series to obtain

$$\frac{d}{dt}\begin{pmatrix}\mathrm{u}\\ \mathrm{v}\end{pmatrix} = \begin{pmatrix}\mathrm{F}(\mathrm{u},\mathrm{v},\phi(\mathrm{v}))\\ \mathrm{G}(\mathrm{u},\mathrm{v},\phi(\mathrm{v}))\end{pmatrix} + \begin{pmatrix}\partial_\phi \mathrm{F}(\mathrm{u},\mathrm{v},\phi(\mathrm{v}))\\ \partial_\phi \mathrm{G}(\mathrm{u},\mathrm{v},\phi(\mathrm{v}))\end{pmatrix}\delta\phi, \tag{7.19}$$

where ∂_ϕ denotes the partial derivative $\partial/\partial\phi$, and ϕ is set to $\phi(\mathrm{v})$ after the derivative is calculated. Perturbation of the volume fraction of polymer, $\delta\phi$, can be determined using eqn (7.10) and (7.11) for the gel swelling as described below. The perturbation $\delta\phi$ is assumed to result from the oscillatory part of the stimulus, whereas the static mechanical action is accounted for in the free-running solution and does not need to be weak.

For the force-controlled stimuli, $\delta\phi$ is found in the form

$$\delta\phi = c_0 K(\mathrm{v})^{-1}\delta F, \tag{7.20}$$

where the function $K(\mathrm{v})$ depends on the partial concentration of oxidized catalyst v (see eqn (7.21)). The effect of the force-controlled deformation, $F = F_0 + \delta F$, on the volume fraction of polymer BZ gel dynamics is captured by eqn (7.20) where $K(\mathrm{v})$ is

$$K(\mathrm{v}) = [(3\lambda^2 + 2\lambda F_0)\partial_\phi P(\phi,\mathrm{v}) - c_0\phi_0(\lambda\phi)^{-2}]|_{\phi=\phi(\mathrm{v}),\ \lambda=\lambda(\mathrm{v})} \tag{7.21}$$

In the above equation, $P(\phi,\mathrm{v}) = \pi_{FH}(\phi) + \chi^*\mathrm{v}\phi^2 + c_0\phi(2\phi_0)^{-1}$, the functions $\phi(\mathrm{v})$ and $\lambda(\mathrm{v})$ are the solutions of eqn (7.10) and (7.11) that can be written as $\phi = \phi_0\lambda^{-2}(\lambda + F_0)^{-1}$ and $c_0\lambda^{-1} = P(\phi,\mathrm{v})$, respectively.

Eqn (7.19)–(7.21) provide the explicit form of eqn (7.12) for the force-controlled mechanical action imposed on the BZ gel. To determine the PRC for the system, we also need the Jacobian matrix for a free-running system (see eqn (7.14)). The Jacobian matrix is calculated as

$$D\{\mathbf{f}(\overline{\mathbf{x}})\} = \begin{pmatrix} \partial_u \mathrm{F} & \partial_v \mathrm{F} \\ \partial_u \mathrm{G} & \partial_v \mathrm{G} \end{pmatrix} - \chi^*\phi^2 M \begin{pmatrix} 0 & \partial_\phi \mathrm{F} \\ 0 & \partial_\phi \mathrm{G} \end{pmatrix} \tag{7.22}$$

The function M in eqn (7.22) is defined by the following equation

$$M = (3\lambda^2 + 2\lambda F_0)K^{-1}, \tag{7.23}$$

where K is given by eqn (7.21). The Jacobian matrix, eqn (7.22), is calculated along the limit cycle of the free-running system determined by eqn (7.19) at $\delta\phi = 0$. Hence, the volume fraction of polymer ϕ and the factor M in eqn (7.22) are functions of the partial concentration of oxidized catalyst v along the limit cycle (see discussion of eqn (7.21)).

7.3.2 Entraining the Responsive BZ Gel

Eqn (7.18)–(7.22) allow us to formulate the phase dynamics expression, eqn (7.15), which describes the process of the entrainment of the BZ gel dynamics to the mechanical action. Now we demonstrate how the procedure works by considering the entrainment of the gel to an applied force, $F(t) = F_0 + \delta F \sin(2\pi\nu t)$, at $F_0 = \delta F = 0.5$.

First, we find the limit cycle of the free-running system by solving numerically the system of equations for the BZ gel dynamics, eqn (7.18), and for the gel swelling, eqn (7.10) and (7.11), at $F_0 = 0.5$ and $\delta F = 0$. Figure 7.2a shows one cycle of oscillation of the concentrations u and v plotted as functions of t/T, where T is the period of oscillation. To use the definition of phase

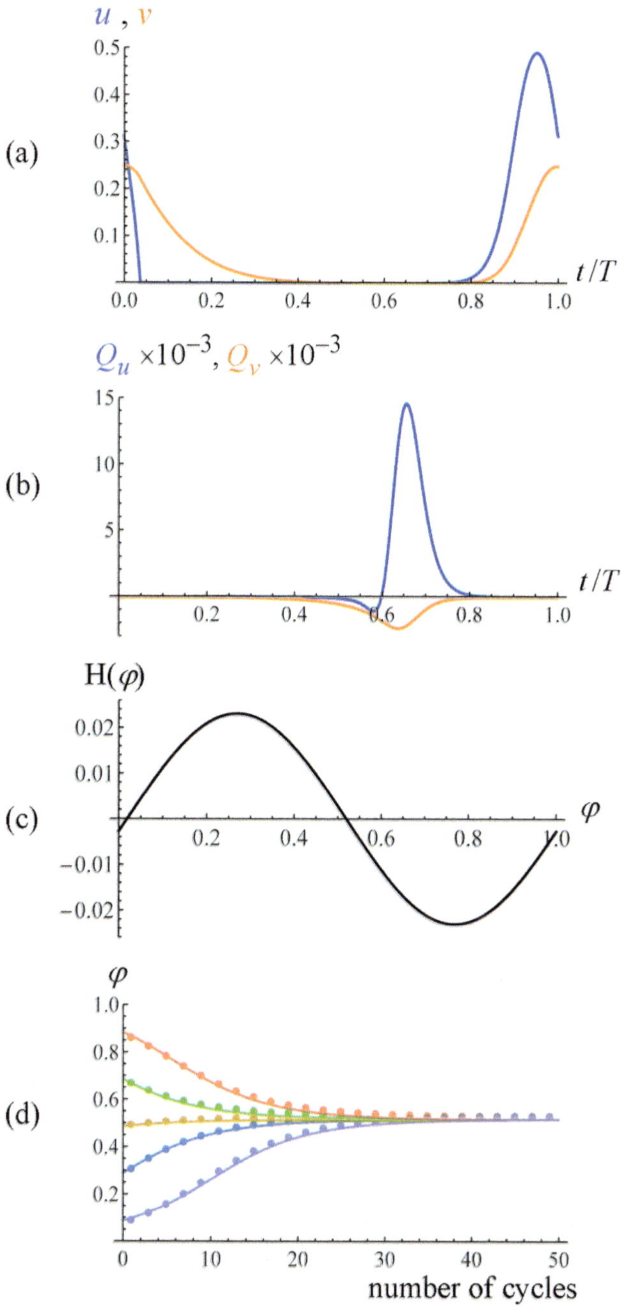

Figure 7.2 Phase dynamics analysis. (a) One cycle of oscillation of the dimensionless concentrations of the activator u (blue) and oxidized catalyst v (orange) in a BZ gel with no periodic forcing. (b) The phase response curves Q_u (blue) and Q_v (orange) obtained using Malkin's method, eqn (7.14). (c) The rate of phase deviation, $H(\varphi)$, obtained by the time averaging, eqn (7.16); φ is the phase difference between oscillation in the BZ gel and oscillating force. (d) Comparison between the kinetics of entrainment obtained by solving the phase dynamics equation (lines) and the full set of equations for the BZ gel under periodic forcing (symbols).

difference shown in Figure 7.1c, the phase of oscillation is chosen such that $\varphi = 0$ corresponds to a moment of time when the concentration of oxidized catalyst v exhibits a maximum.

After obtaining the limit cycle solution for the free-running problem, we use eqn (7.21)–(7.23) to calculate the Jacobian matrix along the limit cycle. Then, we determine the PRC functions Q_u and Q_v by performing numerical integration of eqn (7.14) in the backward direction as described in ref. 18. The resulting PRC functions Q_u and Q_v are shown in Figure 7.2b. The PRC functions are greatest and, therefore, the system is most sensitive to external actions within a narrow range of phases in the second half of the cycle, *i.e.* right before the onset of the next oscillation cycle.

Having calculated the PRC functions Q_u and Q_v, we assume that the frequencies of force oscillation and of free-running BZ-gel oscillation are equal, $\nu = T^{-1}$, and apply the time averaging procedure, eqn (7.16), to determine the function $H(\varphi)$, which governs the slow phase evolution, eqn (7.15). In eqn (7.16), the vector function $\mathbf{g}(\bar{\mathbf{x}})$ within the integrand is defined as

$$\mathbf{g}(\bar{\mathbf{x}}) = c_0 K^{-1} \begin{pmatrix} \partial_\phi \mathrm{F} \\ \partial_\phi \mathrm{G} \end{pmatrix},$$

and is calculated along the limit cycle (see eqn (7.12), (7.19) and (7.20)). The periodic input function $p(t)$ is the oscillatory part of the applied force, so it is defined as $p(t) = 0.5\sin(2\pi\nu t + \pi/2)$ in eqn (7.16). To comply with the definition of phase difference in Figure 7.1c, the input function is calculated with a phase shift of $\pi/2$ relative to the force. The resulting function, $H(\varphi)$ (see Figure 7.2c), has a notable sinusoidal shape, and exhibits one zero at $\varphi \approx 0.52$ that has a negative slope and, hence, corresponds to the state of stable[16–18] entrainment.

Finally, we demonstrate that the phase dynamics in the course of entrainment obtained by solving the full set of equations (*i.e.* eqn (7.1), (7.2), (7.10) and (7.11)) and by solving the equation of phase dynamics, eqn (7.15), are in a good agreement. Figure 7.2d shows the evolution of the phase difference starting from five different initial values obtained from the full set of equations (data points) and from the phase dynamics equation (plotted lines). It is evident that predictions of the phase dynamics equation are quantitatively correct.

Having discussed the implementation of the phase dynamics approach, we use this method to study entrainment of the BZ gel oscillations to periodic compressive force. The force acting on the BZ gel is taken in the form $F(t) = F_0 + \delta F \sin(2\pi\nu t)$, where the force oscillation amplitude is fixed at $\delta F = 0.5$, and the static component of the force is varied at $F_0 = 1$, 2 and 3. The entraining effect of the imposed force is considered for the responsive BZ gel characterized by the coupling parameter $\chi^* = 0.105$. At $\chi^* > 0$, oxidation/reduction of the catalyst cause swelling/de-swelling of the gel in the course of the BZ reaction. The case of non-responsive BZ gel ($\chi^* = 0$) is discussed in ref. 15.

Figure 7.3a shows the phase deviation rate function $H(\varphi)$ obtained for the responsive BZ gel at equal frequencies of the BZ gel and strain oscillations. The phase dynamics is described by the equation $\mathrm{d}\varphi/\mathrm{d}t = H(\varphi)$ according to eqn (7.15). An increase in F_0 applied to the responsive BZ gel (Figure 7.3a) shifts the phase difference in the entrained state, *i.e.* changes the values of φ where $H(\varphi) = 0$ and slope is negative (see also the discussion after eqn (7.17)).

The minimum and maximum of $H(\varphi)$ determine the maximal detuning $\Delta\nu$ at which the strain oscillations are capable of entraining the BZ gel dynamics; this is because at $\Delta\nu \neq 0$, the phase dynamics is described by eqn (7.17), $\mathrm{d}\varphi/\mathrm{d}t = \Delta\nu + H(\varphi)$. (Entrainment is possible if $\mathrm{d}\varphi/\mathrm{d}t = \Delta\nu + H(\varphi) = 0$ has a solution that has a negative slope; therefore, if $\Delta\nu$ lies within the vertical span of H, there is a solution to the latter equation.) Figure 7.3a shows that an increase in the static compression load from $F_0 = 1$ to $F_0 = 3$ slightly decreases the vertical span of $H(\varphi)$ and, thus, narrows the range of frequencies of the force oscillation suitable for entraining the BZ gel.

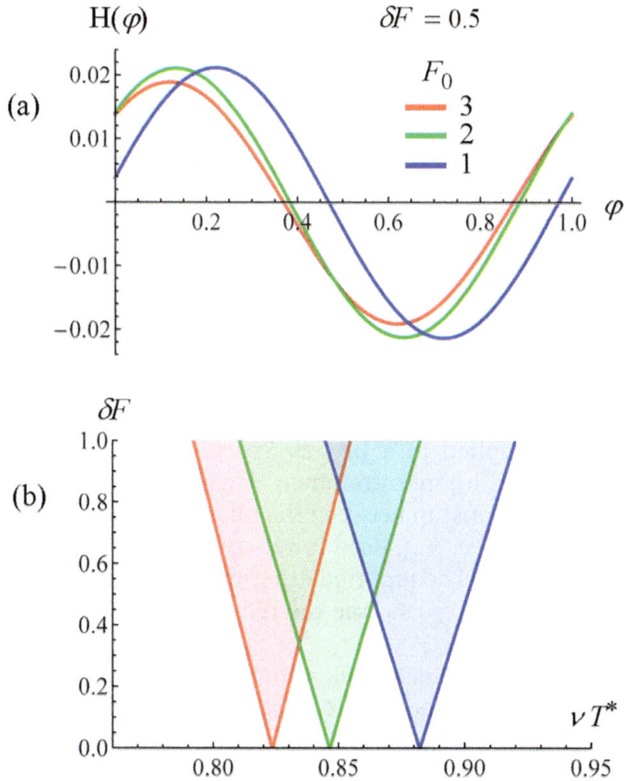

Figure 7.3 Entraining the responsive BZ gel ($\chi^* = 0.105$) by force oscillations at the static compressions of $F_0 = 1$ (blue), 2 (green) and 3 (red). (a) The rate of phase deviation $H(\varphi)$ at $\delta F = 0.5$. (b) Arnold tongues. In (b), νT^* is the dimensionless frequency of force oscillations where T^* is the period of unrestricted oscillation of the BZ gel.

At a given static compression load F_0, entrainment can be also facilitated by an increase in the force oscillation amplitude δF because within the phase dynamics approach, the rate function $H(\varphi)$ is proportional to δF (see eqn (7.16) and recall that $p(t)$ is proportional to δF). With the help of $H(\varphi)$, we plot Figure 7.3b, which shows the entrainment domains mapped on the frequency-amplitude coordinate plane. The entrainment is accessible within the wedge-shaped shaded areas known as Arnold tongues,[16–18] which are plotted for the values of static compression load F_0 = 1, 2 and 3. If the force oscillation amplitude δF is small, entrainment exists only if the frequency of force oscillation is very close to that of the free-running BZ gel (wedge tip). An increase in the force oscillation amplitude δF enables entrainment within an increasingly wider range of the force oscillation frequencies ν (Figure 7.3b). It is also seen in Figure 7.3b that the entrainment domain decreases with an increase in F_0, but the latter effect is rather weak.

The striking feature of the phase deviation rate functions $H(\varphi)$ shown in Figure 7.3 is their sinusoidal shape. It is surprising that the phase dynamics of the highly non-linear oscillations of the BZ gels under external periodic deformations can be adequately described by this simple function. In effect, the phase dynamics is reduced to the Kuramoto model of phase oscillators.[16–18] We now show that $H(\varphi)$ exhibits a sinusoidal shape only for a purely harmonic external input. In addition to the sinusoidal input, we consider the non-linear inputs having the triangular, square, and saw-tooth waveforms shown in Figure 7.4a, and obtain the corresponding rate functions at $\Delta\nu$ = 0 for the force-controlled deformations shown in Figure 7.4b, respectively. Notably, $H(\varphi)$ for the non-linear input functions are no longer sinusoidal, although the rate of change in the phase differences for the entrained state are close for the sinusoidal, triangular and square wave forms.

Our results on the force-controlled entrainment are in a qualitative agreement with experimental results reported in ref. 19, where a square wave-shaped force was applied to a BZ gel. Specifically, the computed phase difference established upon entrainment is close to that observed in the experiment. The relationship between the phase difference φ in Figure 7.1c and the phase difference, φ_{exp}, used in ref. 19 (see Figure 7.3a in the latter paper) is $\varphi = 1.25 - \varphi_{exp}$. The reported experimental value is $\varphi_{exp} \approx 0.8$,[19] and the computed value of $\varphi \approx 0.43$ (see the red line in Figure 7.4b) is close to $0.45 = 1.25 - \varphi_{exp}$.

Finally, the obtained PRCs $\mathbf{Q}(\theta)$, which characterize the effect of an external action on the phase of the BZ gel oscillation, can be employed in solving a range of problems. In particular, these calculations for the entrainment of the BZ gel by a periodic mechanical action of a commensurate frequency can be extended to examine more complex systems. For example, with only a slight modification of the averaging procedure (eqn (7.16)), our approach can be used to study the so-called "m:n" entrainment,[17] which is when the ratio of frequencies of the free-running oscillator and of the external action form a rational number. In the next section,

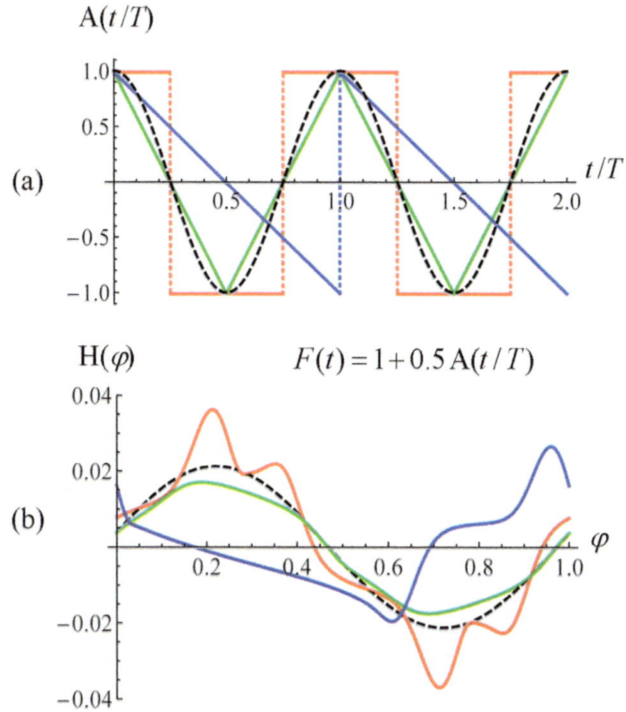

Figure 7.4 Effect of the waveform of imposed oscillation on the rate of phase deviation $H(\varphi)$. (a) Waveform for harmonic (black dashed), square (red), triangular (green) and saw-tooth (blue) oscillations. (b) The rate function $H(\varphi)$ obtained for the force-controlled deformations of the BZ gel. Line colors in (b) correspond to those in (a).

we use our approach to examine synchronization in a system of electrically connected micro-electro-mechanical units made of the self-oscillating BZ gels coupled to piezoelectric films.[20]

7.4 Self-oscillating Gels Coupled Through Piezoelectric Films

Here, we discuss our recently developed model for coupling BZ gels to a piezoelectric (PZ) micro-electro-mechanical system (MEMS).[20] That coupling allows signals to be propagated over long distances, permitting information to be transmitted quickly through remote, non-diffusively coupled oscillations. Importantly, by decoupling the electrical connections from diffusion, the oscillators can be structured into a wide range of topologies. Using our model, we predict the synchronization behavior in these BZ-PZ networks, which could be used for oscillatory computing, ultimately enabling "materials that compute".[20]

Fueled by the internalized BZ reaction, the gels oscillate autonomously in size. With multiple BZ gels, the coherence between these oscillations could be used to perform spatio-temporal processing and recognition tasks.[21–23] Furthermore, BZ gels are pressure-sensitive[24,25] and, thus, they hold the potential to respond to human touch and other physical actions. We find it useful to think of this BZ-PZ hybrid in terms of a cellular structure, in which each cell contains a gel-piezoelectric unit. The simplest coupling between "cells" is depicted in Figure 7.5a. The expansion of the BZ gel in one unit deflects the PZ cantilever. That deflection generates a voltage that causes a deflection in the second cantilever, which in turn imposes a strain on the underlying BZ gel. The oscillations of the second BZ gel are

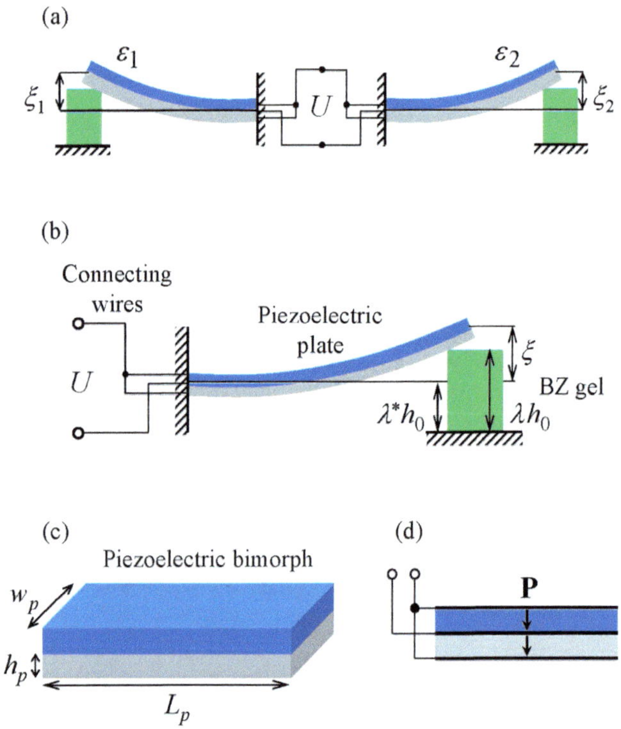

Figure 7.5 Electrically coupled piezoelectric MEMS actuated by self-oscillating polymer gels. (a) Two gel-piezoelectric units coupled through the parallel electric connection. (b) The deflection of piezoelectric cantilever ξ and the electric potential difference U caused by the swollen gel having the degree of swelling λ; h_0 is the size of undeformed cube-shaped gel. (c) The piezoelectric cantilever consists of two layers having the length L_p, width w_p and layer thickness h_p. (d) The cantilever is fabricated from a polarized piezoelectric material (polarization **P**); the internal and external surface electrodes are connected in parallel. In calculations, the dimensions are taken to be $h_0 = 0.5$ mm, $L_p = w_p = 1$ mm, and $h_p = 10$ μm.

modified, and continued coupling leads to synchronization of the two cells.

The oscillations of the BZ gels are driven by the chemical energy stored in the BZ reaction, and so no external energy input is required to couple the BZ-PZ cells. Similarly, there is no amplification apparatus or external information processing required; the system is acting autonomously and independently. This behavior is therefore distinct from pulse-coupled chemical reactors,[23] which interact through computerized electric pumps and valves with a system of light-sensitive BZ beads that are organized into a virtual network by a computer algorithm[26] or the electrically coupled continuously stirred tank reactors[27,28] that require signal amplification. The self-contained and autonomous function of the miniature BZ-PZ cells creates opportunities to integrate them into a larger material's "fabric".

The size scale of the basic cell units is critical to their function (Figure 7.5b). For example, the gel in a basic cell unit must be small enough to exhibit spatially uniform oscillations, and the cantilever must be thin (and compliant) enough that it is readily deflected by the low forces generated by the expanding gel. The cantilever's parallel bimorph structure consists of two identical layers of a polarized piezoelectric material (Figure 7.5c and d). The internal and external surface electrodes are connected in parallel (Figure 7.5d).

To capture the dynamics of this basic unit, we couple equations for the volumetric oscillations of the small-sized BZ gel and bending elasticity of the piezoelectric bimorph beam. As in the previous section, we assume that the gel is formed from PNIPAAm chains containing grafted ruthenium metal-ion catalysts[3] and use eqn (7.1)–(7.4) to describe the kinetics of the BZ reaction in terms of the concentrations of activator, u, oxidized catalyst, v, and volume fraction of polymer, ϕ. Periodic variations in v due to the BZ reaction affect the polymer–solvent interactions and drive the gel's rhythmic expansion and contraction. The allowable dimensions of the sample depend on the elasticity of the network, the force exerted by the cantilever F_g and the osmotic pressure of the polymer. For simplicity, we assume that the gel exhibits uniaxial deformations, which could be achieved by confining the gel within a capillary, so that the degree of swelling in the transversal direction, $\lambda_\perp$, is constant. Then, the gel's degree of swelling in the longitudinal direction, λ, as a function of v and F_g is provided by eqn (7.8), because the volume fraction of polymer depends on λ and $\lambda_\perp$ as $\phi = \phi_0\lambda^{-1}\lambda_\perp^{-2}$.

The composition of the BZ substrate is the same as in the previous section so in calculations, the corresponding Oregonator model parameters are $\varepsilon_{BZ} = 0.212$, $q = 9.52 \times 10^{-5}$ and $\mu_{BZ} \approx 7 \times 10^2$. The stoichiometric parameter f and the catalyst content c_{Ru} are assumed to be $f = 1$ and $c_{Ru} = 3$. We consider cases where the volume fraction of polymer in the undeformed BZ gel is $\phi_0 = 0.139$ or $\phi_0 = 0.16$, and the cross-link density in the gel is $c_0 = 4 \times 10^{-4}$ or $c_0 = 4.5 \times 10^{-4}$. As above, the polymer–solvent interaction is set to the value for the PNIPAAm-water interaction at 20 °C, the interaction parameter describing the hydrating effects of the oxidized catalyst is $\chi^* = 0.105$ and the undeformed gel size is $h_0 = 0.5$ mm.

The overall behavior of the PZ component in this system is captured by quasi-static equations, arising from the relatively small frequency of the chemo-mechanical oscillations *vis-à-vis* the eigenfrequency of the cantilever. These equations[29] provide the deflection, ξ, and the electric charge, Q, of the bimorph as linear functions of the force, F, applied to the cantilever and the electric potential difference (voltage), U, between the electrodes (Figure 7.5b):

$$\xi = m_{11}F + \varepsilon m_{12}U \tag{7.24}$$

$$Q = \varepsilon m_{12}F + m_{22}U \tag{7.25}$$

The force F is applied to the cantilever's tip in the vertical direction. The coefficients m_{11}, m_{12} and m_{22} depend on the cantilever dimensions and material's properties of the piezoelectric:

$$m_{11} = \frac{\left(L_\mathrm{p}/h_\mathrm{p}\right)^3}{2w_\mathrm{p}E} \tag{7.26}$$

$$m_{12} = \frac{3}{4}d_{31}\left(L_\mathrm{p}/h_\mathrm{p}\right)^2 \tag{7.27}$$

$$m_{22} = 2w_\mathrm{p}\varepsilon_{33}(1 - k^2/4)(L_\mathrm{p}/h_\mathrm{p}) \tag{7.28}$$

Here, L_p, w_p and h_p are the respective length, width and layer thickness of the piezoelectric bimorph plate (see Figure 7.5c); E, d_{31} and ε_{33} are the Young's modulus, piezoelectric constant and dielectric constant of the piezoelectric material, respectively; and $k = (d_{31}^2E/\varepsilon_{33})^{1/2}$ is the electromechanical coupling factor characterizing the piezoelectric material. The force polarity, ε, on the right-hand side of eqn (7.24) and (7.25) is either +1 when the direction of polarization of the piezoelectric material (Figure 7.5d) and that of the applied force coincide or −1 when they are in opposite directions.

For shorted electrodes ($U = 0$), eqn (7.24) gives Hooke's law, $F^{(0)}(\xi) = m_{11}^{-1}\xi$. In the open circuit configuration, where $Q = 0$, the piezoelectric effect impacts the bending rigidity of the cantilever, so that $F(\xi) = (1 - \zeta^2)^{-1}F^{(0)}(\xi)$, where $\zeta^2 = m_{12}^2(m_{11}m_{22})^{-1}$ The value of ζ does not depend on the bimorph dimensions, as follows from eqn (7.26)–(7.28); namely, $\zeta = 3k(16 - 4k^2)^{-1/2}$, where k is the electromechanical coupling factor introduced in the previous paragraph. The effect of piezoelectricity on the bending rigidity is small as $\zeta^2 \ll 1$ for a typical piezoelectric. The cantilever is assumed to be made of polarized Lead-Zirconate-Titanate ceramics (PZT), one of the most commonly used piezoelectrics.[30] If obtained under common processing conditions, a PZT ceramic has a Young's modulus of $E = 50$ GPa, and typical values for the other parameters are $d_{31} = -1.5 \times 10^{-10}$ m V^{-1} and $\varepsilon_{33} = 1.8 \times 10^3\varepsilon_0$, where $\varepsilon_0 = 8.85$ pF V^{-1} is the dielectric constant of vacuum,[30] so that $k \approx 0.268$ and $\zeta \approx 0.201$.

Alternative processing methods[31,32] can yield a two-fold increase in d_{31}, and hence, a four-fold increase in ζ^2. In calculations, we set $d_{31} = -3 \times 10^{-10}$m V^{-1} (*i.e.* twice the typical value) assuming that PZT is fabricated through advanced processing methods, and the dimensionless values characterizing the piezoelectric material are $k \approx 0.531$ and $\zeta \approx 0.413$. Finally, the piezoelectric bimorph plate dimensions are taken to be $L_p = w_p = 1$ mm, $h_p = 10$ μm.

The effect of the cantilever on the BZ gel is a result of the force F_g applied to the gel surface, and so once the bending force $F(\xi)$ is determined through eqn (7.24) and (7.25), F_g as a function of λ is known, since $F_g = F(\xi)$, and the cantilever deflection is caused by the gel swelling, so $\xi = (\lambda - \lambda^*)h_0$ (Figure 7.5b). Hence, the dynamics of the basic unit is described by eqn (7.1)–(7.4) and (7.8), with the force F_g depending on the degree of swelling λ.

There are two primary modes of coupling between individual BZ-PZ cells: serial (Figure 7.6a) and parallel (Figure 7.6b). The connected units are assumed to be identical, but can differ in their force polarity. The oscillatory dynamics of each unit is described by eqn (7.1)–(7.4) and (7.8) but the force acting on a given gel now depends on the degrees of swelling of all gels in the system. To obtain the equations for forces in a system of n units, we rewrite eqn (7.24) and (7.25) as

$$F_i = F_i^{(0)} - m_{11}^{-1} m_{12} \varepsilon_i U_i \quad (7.29)$$

$$Q_i = m_{12} \varepsilon_i F_i^{(0)} + m_{22}(1 - \zeta^2) U_i \quad (7.30)$$

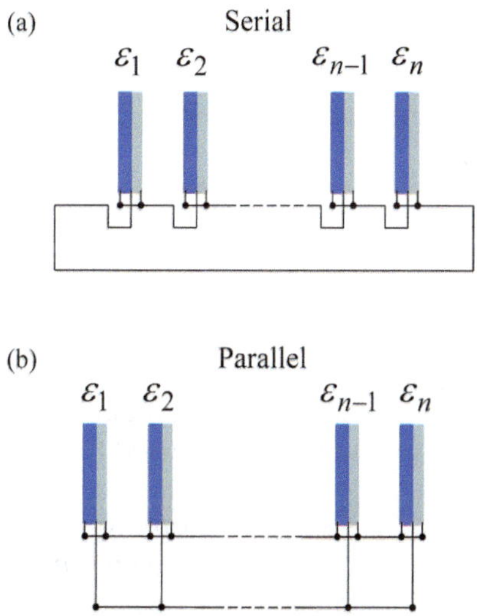

Figure 7.6 Multiple gel-piezoelectric units connected to circuits. (a) Serial connection. (b) Parallel connection. The units are identical and can differ only in their force polarity $\varepsilon_i = \pm 1$, where $1 \le i \le n$ is the unit number. The two parts of a bimorph piezoelectric plate are distinguished by coloring in blue and gray.

where, $i = 1,2,\ldots,n$ labels the units. In the above equations, $F_i^{(0)} = m_{11}^{-1}\xi_i$ is Hooke's law for an individual bending plate without accounting for the piezoelectric effect, and $\zeta^2 = m_{12}^2(m_{11}m_{22})^{-1}$ (see the discussion after eqn (7.28)). If the units are connected in series (Figure 7.6a), they all have the same electric charge, $Q_i = Q$, and the net electric potential difference is zero, $\sum_{i=1}^{n} U_i = 0$. Summation of eqn (7.30) for all $1 \le i \le n$ gives $Q = m_{12}n^{-1}\sum_{j=1}^{n} \varepsilon_j F_j^{(0)}$, so the potential difference on the ith unit is found from eqn (7.30) to be

$$U_i = -m_{12}m_{22}^{-1}\left(1-\zeta^2\right)^{-1}\left[\varepsilon_i F_i^{(0)} - n^{-1}\sum_{j=1}^{n}\varepsilon_j F_j^{(0)}\right] \tag{7.31}$$

After substituting eqn (7.31) into eqn (7.29), we obtain the following equation for the bending force acting on the ith cantilever:

$$F_i = F_i^{(0)} + \kappa\left[F_i^{(0)} - \varepsilon_i n^{-1}\sum_{j=1}^{n}\varepsilon_j F_j^{(0)}\right], \tag{7.32}$$

where $\kappa = \zeta^2(1-\zeta^2)^{-1}$. For the parallel connection, there is no net electric charge in the system, $\sum_{i=1}^{n} Q_i = 0$, and all bimorphs have the same voltage, $U_i = U$. Summation of eqn (7.20) yields

$$U = -m_{12}m_{22}^{-1}\left(1-\zeta^2\right)^{-1} n^{-1}\sum_{j=1}^{n}\varepsilon_j F_j^{(0)} \tag{7.33}$$

Hence, the bending force found from eqn (7.29) is:

$$F_i = F_i^{(0)} + \kappa\varepsilon_i n^{-1}\sum_{j=1}^{n}\varepsilon_j F_j^{(0)} \tag{7.34}$$

Given eqn (7.32) and (7.34), the forces acting on the ith gel can be found as described above, and the latter forces depend on the degrees of swelling of all gels. Importantly, however, the cross-terms depicting the interaction between the gel-piezoelectric units are small due to the weak interaction, as κ is small. At the model parameter values used for calculations, we obtain $\kappa \approx 0.206$.

An initial example is the electrical coupling of two BZ-PZ cells (Figure 7.5a). If one solves the equations that govern coupling, it becomes clear that the intercellular interactions must lead to synchronized chemo-mechanical oscillations in the cells, regardless of their initial oscillatory state. In addition, the mode of synchronization, *i.e.* the phase difference of oscillations in the synchronized state, depends on the force polarity of the two units. The interacting units exhibit anti-phase synchronization at $\varepsilon_1 = \varepsilon_2 = 1$ (Figure 7.7a) or in-phase synchronization if $\varepsilon_1 = 1$ and $\varepsilon_2 = -1$ (Figure 7.7b).

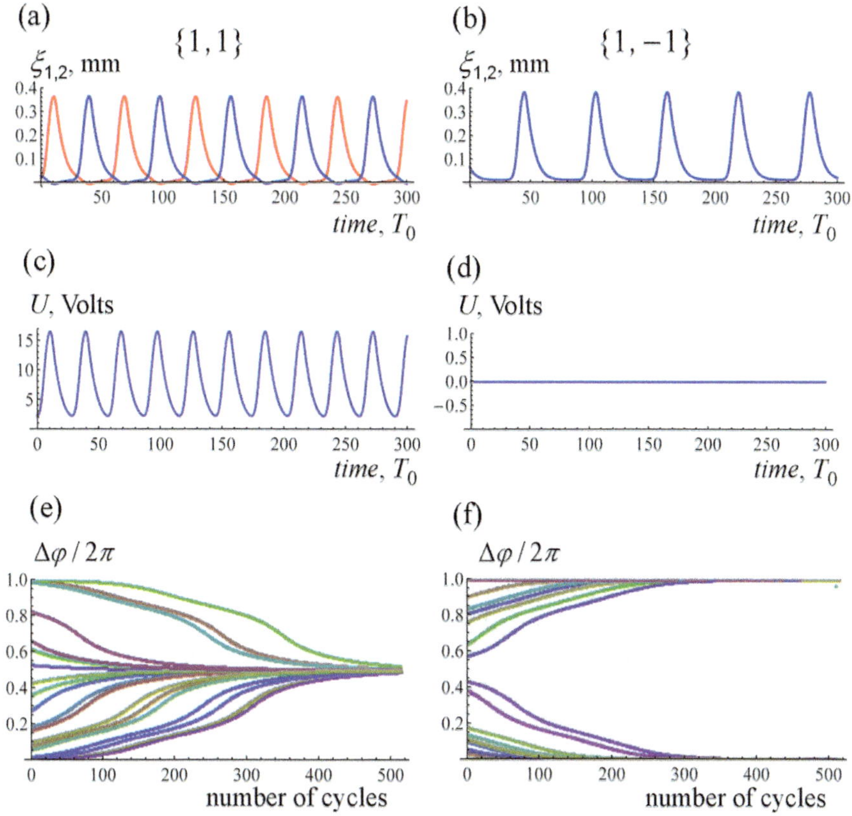

Figure 7.7 Modes of synchronization in the system of two gel-piezoelectric units. Units are connected in parallel as shown in Figure 7.5a. At the force polarity set $\{\varepsilon_i\} = \{1,1\}$, (a) the deflections ξ_1 and ξ_2 oscillate anti-phase, that results in (c) the oscillatory voltage U. At $\{\varepsilon_i\} = \{1,-1\}$, (b) the cantilevers bend in-phase, and (d) no voltage is generated, $U = 0$. No other modes of synchronization are observed in 20 independent simulations starting from random initial values: the phase difference $\Delta\varphi$ converges to (e) π at $\{\varepsilon_i\} = \{1,1\}$, and to (f) 0 or 2π at $\{\varepsilon_i\} = \{1,-1\}$. The calculations were performed at the volume fraction of polymer and the cross-link density in the undeformed BZ gel of $\phi_0 = 0.139$ and $c_0 = 4.5 \times 10^{-4}$, respectively, and at the values of $\lambda_\perp$ and λ^* (see Figure 7.5b) of $\lambda_\perp = 1.4$ and $\lambda^* = 1.55$. All other model parameters were the same as described in the text.

This creates an opportunity for further information transfer, in that the two synchronization modes can be distinguished through the resulting patterns of the voltage U created by the piezoelectric bimorphs. It can be shown that $U = -m_{12}(2m_{11}m_{22})^{-1}(1 - \zeta^2)^{-1}(\xi_1 \pm \xi_2)$, where ξ_1 and ξ_2 are the deflections of the two cantilevers, and the (+) and (−) signs in the parentheses correspond to the force polarity sets {1,1} and {1,−1}, respectively. The latter equation for the voltage U follows directly from eqn (7.31) and (7.33). Hence, the anti-phase gel oscillations at $\{\varepsilon_i\} = \{1,1\}$ generate oscillatory voltage (Figure 7.7c), whereas $U = 0$ for the in-phase oscillation at $\{\varepsilon_i\} = \{1,-1\}$ (Figure 7.7d). The difference between the waveforms in Figure 7.7a and b stems from the various contributions of the piezoelectric effect to the gel–gel interaction for the different sets of ε_i.[20]

The synchronized oscillations in Figure 7.7a and b are the only synchronization modes observed in the simulations at the respective force polarity sets. Figure 7.7e and f show how the phase difference $\Delta\varphi$ evolves in 20 independent simulations starting from different, random individual values of $\Delta\varphi$ for the polarity sets $\{1,1\}$ and $\{1,-1\}$, respectively. In all simulations, $\Delta\varphi$ converges to π in Figure 7.7e, and to 0 or 2π in Figure 7.7f. Figure 7.7e and f also reveal that synchronization can be quite slow, taking up to 300 or 400 cycles. The time to synchronize depends on the initial conditions, but it is fundamentally limited by the coupling, as quantified by the (often small) coupling parameter κ (see eqn (7.32) and (7.34)).

In order to approach the problem of determining the synchronization behavior of n electrically connected gel-piezoelectric units, we take into account the weak coupling between these units and formulate the phase dynamics[16–18] equations by following the same steps as outlined in the previous section (see ref. 20 for details). As a result, the dynamics of the system is described in terms of time-dependent deviation of phase in each oscillator, $\varphi_i(t)$. For n interacting gel-piezoelectric units, the evolution of the difference of phases $\varphi_{ij} = \varphi_i - \varphi_j$ of oscillators i and j is given by the following equation:

$$\mathrm{d}\varphi_{ij}/\mathrm{d}t = \mp\kappa n^{-1}\left[\varepsilon_i \sum_{k=1,k\neq i}^{n} \varepsilon_k H(\varphi_{ki}) - \varepsilon_j \sum_{l=1,l\neq j}^{n} \varepsilon_l H(\varphi_{lj})\right] \tag{7.35}$$

The signs (−) and (+) on the r.h.s. of eqn (7.35) respectively depict the unit connections in series and parallel. The function $H(\varphi)$, which is calculated numerically (see ref. 20), corresponds to the rate of the phase shift of an individual BZ gel as a result of the interaction with another gel at the phase difference between the two oscillators of φ. When applied to a system of weakly interacting oscillators, $H(\varphi)$ is usually called the *connection* function. The numerically obtained connection function $H(\varphi)$ is plotted in Figure 7.8a. The phase normalization is such that $0 \leq \varphi \leq 1$. Figure 7.8a shows that interactions can cause positive or negative phase shifts depending on the phase difference.

Using eqn (7.35), we start by showing that the synchronization modes in Figure 7.7 are the sole ones possible in the system of two gel-piezoelectric units. At $n = 2$, eqn (7.35) can be written as $\mathrm{d}\varphi_{21}/\mathrm{d}t = \mp 2^{-1}\kappa\varepsilon_1\varepsilon_2[H(-\varphi_{21}) - H(\varphi_{21})]$. The stable modes of synchronization correspond to those zeroes of the function on the r.h.s. where this function has a negative slope. The function $H(-\varphi) - H(\varphi)$ (Figure 7.8b) has zeroes at $\varphi = 0, 0.5$, and 1. For the two units connected in parallel (Figure 7.5a), $\varphi_{21} = 0.5$ is the stable solution at $\varepsilon_1\varepsilon_2 > 0$ shown in Figure 7.7a, and $\varphi_{21} = 0$ and 1 are the stable solutions at $\varepsilon_1\varepsilon_2 < 0$ shown in Figure 7.7b.

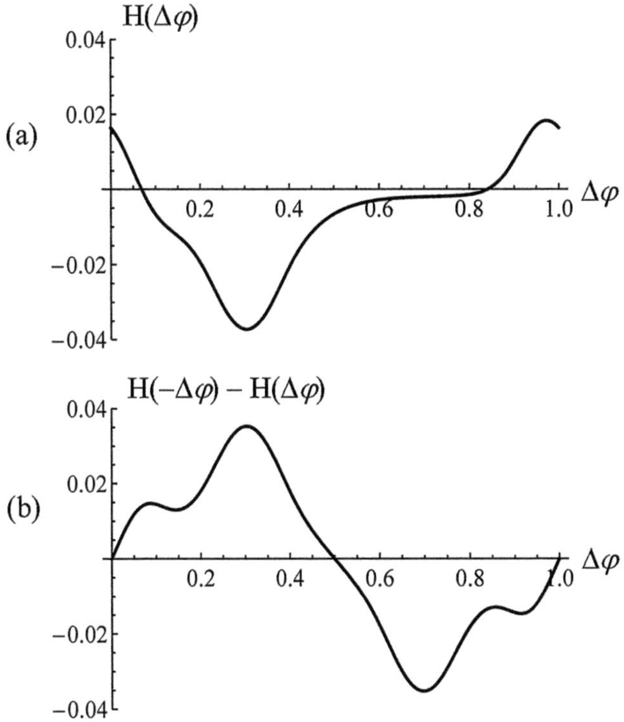

Figure 7.8 Phase dynamics analysis. (a) The connection function $H(\Delta\varphi)$ that characterizes the rate of phase shift of an individual BZ gel due to the interaction with another gel at the phase difference between the two oscillators of $\Delta\varphi$. (b) The function $H(-\Delta\varphi) - H(\Delta\varphi)$ describes synchronization in the system of two units. The calculations were performed at the volume fraction of polymer and the cross-link density in the undeformed BZ gel of $\phi_0 = 0.16$ and $c_0 = 4 \times 10^{-4}$, respectively. The values of $\lambda_\perp$ and λ^* are taken equal to λ_{st}, which corresponds to the steady-state value for an isotropic BZ gel. All other model parameters were the same as described in the text. At the used values of model parameters, $\lambda_\perp = \lambda^* = \lambda_{st} \approx 1.65$.

Certain features of eqn (7.35) allow the description of synchronization in many-cell BZ-PZ hybrids to be simplified. First, the simultaneous change of sign of all the force polarities $\{\varepsilon_{ij}\}$ does not change the phase dynamics, and thus does not affect the synchronization modes. Next, the phase dynamics of units connected in series and in parallel differ only in the sign of the function on the r.h.s. of eqn (7.35). Hence, the steady-states of eqn (7.35), *i.e.* the synchronization modes, are the same for both types of connections. Stability of a steady-state is determined by eigenvalues of the Jacobian matrix, which have opposite signs for the serial and parallel connections. Therefore, as we have noted previously, the synchronization modes that are stable for the serial connection are unstable if the units are connected in parallel, and *vice versa*. Finally, we note that it is possible for modes to display saddle-like stability; such modes would be unstable and therefore ill-suited for both serial and parallel connections of the BZ-PZ unit cells.

The above properties of eqn (7.35) are utilized to determine all possible modes of synchronization for a system containing three gel-piezoelectric units. Only two sets of the force polarity of the units, $\{\varepsilon_i\}$, namely, {1,1,1} and {1,1,−1}, need to be considered. For three units, either all the polarities are identical or only one differs. The respective results are presented in Figure 7.9a and b. The synchronization modes are obtained as the steady-states of two phase difference equations. Precisely, the simultaneous solutions of the equations $d\varphi_{21}/dt = 0$ and $d\varphi_{32}/dt = 0$ are determined. For each of the equations, the solutions form a curve or several curves within the area of the plane with coordinates $0 \le \varphi_{21} \le 1$ and $0 \le \varphi_{32} \le 1$. The intersections of the two sets of curves give the synchronization modes. In Figure 7.9a and b, the intersections are marked by colored disks to distinguish between the modes, which are stable for the serial (green) and parallel (red) connections or unstable for any of them (gray). It is possible to generate effectively equivalent modes that result from different combinations of the units or are shifted only by a period of oscillation.

The insets highlight the distinct oscillation patterns for the stable synchronization modes, which are predicted by the phase dynamics equations. There are two distinct modes for both the serial and parallel connections at the force polarity set {1,1,1} (Figure 7.9a) and only one mode for each type of

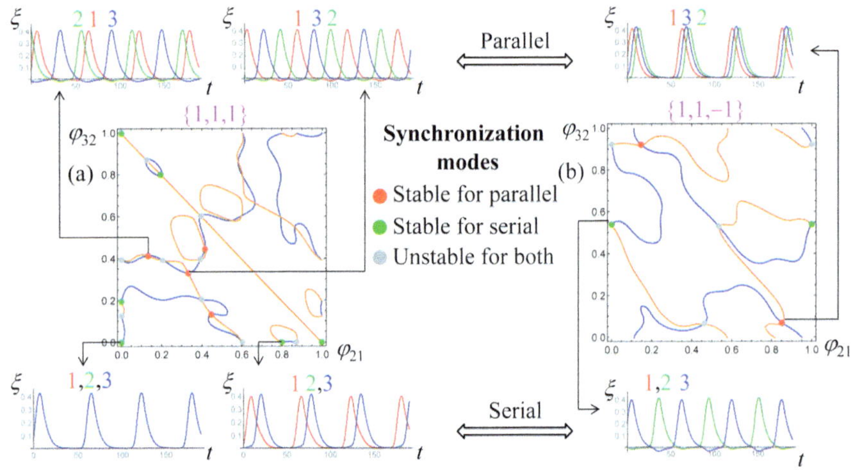

Figure 7.9 Synchronization modes in systems of three gel-piezoelectric units. Obtained by graphical solution of the equations of phase dynamics for the force polarity sets $\{\varepsilon_i\}$ of (a) {1,1,1}, and (b) {1,1,−1}. The green and red disks mark the modes stable for the serial and parallel connections, respectively. The gray disks mark the modes unstable for both types of connection. The insets show the cantilever oscillation patterns for the stable synchronization modes. The cantilever deflection curves for the individual units in the system are labeled by the corresponding unit numbers 1, 2 or 3. The model parameters used are the same as in Figure 7.8.

connection at the polarities of {1,1,−1} (Figure 7.9b). The oscillation patterns were also obtained by solving the full set of equations, eqn (7.1)–(7.4), (7.8) and (7.32) or eqn (7.34), with the initial conditions that correspond to a chosen synchronization mode perturbed by a random phase shift up to 0.01. The latter calculations confirm the results for the stability of the modes obtained from the phase dynamics approach. This new approach lays the groundwork for studies of large n systems in the future.

7.5 Conclusions

In the first part of this study, we modeled the entrainment of self-oscillating BZ gels to externally imposed, time-periodic mechanical deformation controlled by either strain or force. For simplicity, we focused on sufficiently small-sized samples that the chemo-mechanical oscillations could be assumed to be spatially uniform. We demonstrated how the phase dynamics approach can be applied to analyze the behavior of the BZ gels subjected to these modes of deformation. We validated our phase dynamics calculations by comparing the results to those obtained by solving the full set of equations for the BZ gel under periodic forcing. We also validated our approach by comparing our results to the available experimental data.

We then built on this validated model to design BZ-PZ composites that exploit the inherent properties of the individual components to perform sensing, actuation, communication and computation. Notably, the materials perform these functions without external power sources or implanted sensors and microelectronics. In effect, the system is truly biomimetic since it performs these biological-inspired functions in an autonomous manner.

There have been various attempts to create chemical computers[21] that rely on the interactions between traveling chemical waves to perform logic operations. Nevertheless, chemical computing systems are limited by the local coupling of the reacting species due to diffusion as the chemical waves propagate throughout the system. The BZ-PZ described here addresses this critical limitation. In particular, with the introduction of the electrical wires between the units, the chemo-mechano-electrical coupling in the system can be propagated quickly over long distances, and thus permits the synchronization of non-diffusively coupled oscillators. Moreover, the presence of the connecting wires allows the oscillators to be organized into arbitrary topologies.

Thanks to the synchronization between these oscillators, computational tasks such as pattern recognition can be performed with this system. Our new model allows us to predict the synchronization in networks of BZ-PZ oscillators, and hence tailor the computational capabilities of the system. Ultimately, the gel-piezoelectric composite could be placed on top of a porous substrate filled with the reagents for the BZ reaction and the whole system embedded within a polymeric material. Consequently, the system would mimic the gel inserts used in footwear. Pressure from each step would liberate the BZ reagents and fuel the oscillations. The mode of locomotion would impact the synchronization, and therefore the system would be capable of detecting changes in gait . The latter could, for instance, be indicative of changes in health. Broadly speaking, the quest to design smart materials capable of sensing their environment, processing information and reacting to complex stimuli will facilitate the advent of new systems that will interface with humans to offer tactile, temperature and photonic inputs to smart clothing, synthetic skin for robotic hands and even prosthetic devices.

Acknowledgement

Financial support from NSF, through grant number DMR-1344178, is gratefully acknowledged.

References

1. M. A. C. Stuart, W. T. S. Huck, J. Genzer, M. Müller, C. Ober, M. Stamm, G. B. Sukhorukov, I. Szleifer, V. V. Tsukruk, M. Urban, F. Winnik, S. Zauscher, I. Luzinov and S. Minko, *Nat. Mater.*, 2010, **9**, 101.
2. M. A. McEvoy and N. Correll, *Science*, 2015, **347**, 1261689.
3. R. Yoshida, T. Takahashi, T. Yamaguchi and H. Ichijo, *J. Am. Chem. Soc.*, 1996, **118**, 5134.
4. M. Ma, L. Guo, D. G. Anderson and R. Langer, *Science*, 2013, **339**, 186.
5. Q. Sun, W. Seung, B. J. Kim, S. Seo, S. W. Kim and J. H. Cho, *Adv. Mater.*, 2015, **27**, 3411.
6. C. Dagdeviren, Y. Shi, P. Joe, R. Ghaffari, G. Balooch, K. Usgaonkar, O. Gur, P. L. Tran, J. R. Crosby, M. Meyer, Y. Su, R. C. Webb, A. S. Tedesco, M. J. Slepian, Y. Huang and J. A. Rogers, *Nat. Mater.*, 2015, **14**, 728.
7. J. J. Tyson and P. C. Fife, *J. Chem. Phys.*, 1980, **73**, 2224.
8. V. V. Yashin, S. Suzuki, R. Yoshida and A. C. Balazs, *J. Mater. Chem.*, 2012, **22**, 13625.
9. V. V. Yashin, O. Kuksenok and A. C. Balazs, *J. Phys. Chem. B*, 2010, **114**, 6316.
10. V. V. Yashin and A. C. Balazs, *Macromolecules*, 2006, **39**, 2024.
11. V. V. Yashin and A. C. Balazs, *J. Chem. Phys.*, 2007, **126**, 124707.
12. R. J. Atkin and N. Fox, *An Introduction to the Theory of Elasticity*, Longman, New York, 1980.
13. S. Hirotsu, *J. Chem. Phys.*, 1991, **94**, 3949.
14. O. Kuksenok, V. V. Yashin and A. C. Balazs, *Phys. Rev. E*, 2008, **78**, 041406.
15. V. V. Yashin, S. P. Levitan and A. C. Balazs, *Chaos*, 2015, **25**, 064302.
16. A. Pikovsky, M. Rosenblum and J. Kurth, *Synchronization: A Universal Concept in Nonlinear Sciences*, Cambridge University Press, Cambridge, UK, 2003.
17. F. C. Hoppensteadt and E. M. Izhikevich, *Weakly Connected Neural Networks*, Springer, New York, 2007.
18. E. M. Izhikevich, *Dynamical Systems in Neuroscience: The Geometry of Excitability and Bursting*, MIT Press, Cambridge, 2007, Ch. 10.

19. T. Shiota, Y. S. Ikura and S. Nakata, *J. Phys. Chem. B*, 2013, **117**, 2215.
20. V. V. Yashin, S. P. Levitan and A. C. Balazs, *Sci. Rep.*, 2015, **5**, 11577.
21. A. Adamatzky, B. D. L. Costello and T. Asai, *Reaction-diffusion Computers*, Elsevier, Boston, 2005.
22. F. C. Hoppensteadt and E. M. Izhikevich, *IEEE Trans. Neural Netw.*, 2000, **11**, 734.
23. P. l. Gentili, V. Horváth, V. K. Vanag and I. R. Epstein, *Int. J. Unconv. Comput.*, 2012, **2**, 177.
24. I. C. Chen, O. Kuksenok, V. V. Yashin, A. C. Balazs and K. J. Van Vliet, *Adv. Funct. Mater.*, 2012, **22**, 2535.
25. O. Kuksenok, V. V. Yashin and A. C. Balazs, *Soft Matter*, 2007, **3**, 1138.
26. M. R. Tinsley, S. Nkomo and K. Showalter, *Nat. Phys.*, 2012, **8**, 662.
27. W. Hohmann, M. Kraus and F. W. Schneider, *J. Phys. Chem. A*, 1999, **103**, 7606.
28. M. F. Crowley and R. J. Field, *J. Phys. Chem.*, 1986, **90**, 1907.
29. R. G. Ballas, *Piezoelectric Multilayer Beam Bending Actuators: Static and Dynamic Behavior and Aspects of Sensor Integration*, Springer, New York, 2007.
30. A. Preumont, *Mechatronics: Dynamics of Electromechanical and Piezoelectric Systems*, Springer, Dordrecht, 2006.
31. S. H. Baek, J. Park, D. M. Kim, V. A. Aksyuk, R. R. Das, S. D. Bu, D. A. Felker, J. Lettieri, V. Vaithyanathan, S. S. N. Bharadwaja, N. Bassiri-Gharb, Y. B. Chen, H. P. Sun, C. M. Folkman, H. W. Jang, D. J. Kreft, S. K. Streiffer, R. Ramesh, X. Q. Pan, S. Trolier-McKinstry, D. G. Schlom, M. S. Rzchowski, R. H. Blick and C. B. Eom, *Science*, 2011, **334**, 958.
32. S. H. Baek, M. S. Rzchowski and V. A. Aksyuk, *MRS Bull.*, 2012, **37**, 1022.

Subject Index